Event Analytics across Languages and Communities

Ivana Marenzi • Simon Gottschalk •
Eric Müller-Budack • Marko Tadić • Jane Winters
Editors

Event Analytics across Languages and Communities

Springer

Editors
Ivana Marenzi
L3S Research Center
Leibniz University of Hannover
Hannover, Germany

Simon Gottschalk
L3S Research Center
Leibniz University Hannover
Hannover, Germany

Eric Müller-Budack
TIB – Leibniz Information Centre
for Science and Technology
Hannover, Germany

Marko Tadić
Faculty of Humanities and Social Sciences
University of Zagreb
Zagreb, Croatia

Jane Winters
School of Advanced Study
University of London
London, UK

ISBN 978-3-031-64450-4 ISBN 978-3-031-64451-1 (eBook)
https://doi.org/10.1007/978-3-031-64451-1

This Springer imprint is published by the registered company Springer Nature Switzerland AG
The registered company address is: Gewerbestrasse 11, 6330 Cham, Switzerland

Preface

Unexpected incidents such as natural disasters and terrorist attacks, planned events such as football world championships and long-lasting and evolving events such as the migration crisis in Europe and military conflicts affect communities and societies across languages around the globe. News Web sites and social media cover these events, leading to a vast amount of multilingual event information from heterogeneous sources. Dealing with such information calls for methodologies, tools and datasets enabling effective cross-lingual interlinking, verification, contextualisation and analytics of event-centric multilingual information originating from different communities, as well as intuitive ways of interacting with multilingual information. Such technologies are of the utmost importance for various stakeholder groups, including digital humanities researchers, memory institutions, publishers, media monitoring companies and journalists. This book presents interdisciplinary and cross-sectoral research results fostering event analytics across languages and communities.

Figure 1 gives an overview of the CLEOPATRA International Training Network (ITN),[1] a central building block of this book. The project offered a unique interdisciplinary and cross-sectoral research and training programme, which explored how we analyse and understand how the major events that influence and shape our lives and societies play out online. Such events are represented in a wealth of resources in the European languages, including English and German, but also languages like Croatian and Slovene with less resources available. Their analysis and exploration were achieved through various studies, the development of novel methodologies in fields such as data mining and natural language processing (NLP) and the creation of new event-centric datasets aggregated in the Open Event Knowledge Graph (OEKG), a multilingual event-centric knowledge graph that contains more than 1 million events in 15 languages.

The Cleopatra ITN project started in January 2019 and ran until June 2023, so it witnessed the transition of NLP from the first Transformer-based language

[1] https://cleopatra-project.eu/

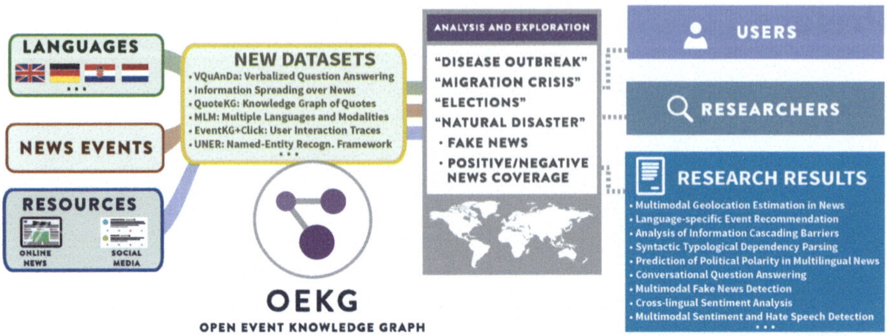

Fig. 1 Overview of the CLEOPATRA ITN and some highlights also covered in this book

models into the era of Large Language Models (LLMs), which led to significant breakthroughs in various fields of artificial intelligence (AI) (Zhou et al. 2023), the most important of which concerned the generation and understanding of text, breaking previously prevalent language barriers. The methodologies and resources presented in this book need to be scrutinised in the context of this transition. The presented methodologies include named entity recognition, conversational question answering and the narrativisation of events, which can also, now and in the future, be targeted using LLMs. With this book, our intention is twofold: firstly, to showcase the state-of-the-art NLP methodologies for event analytics before the introduction of LLMs, which can serve as the benchmarks for future developments in these fields, and, secondly, to emphasise the need for the proposed methodologies and resources even in the era of LLMs, addressing the inherent deficiencies of LLMs when it comes to questions of reliability. In a sensitive field facing societies with different cultures and perceptions, truthfulness is an indispensable dimension that can be accomplished through carefully designed model architectures and symbolic knowledge representations, for example, through event-centric knowledge graphs.

One central aspect of the Cleopatra ITN was the creation of resources supporting event analytics, culminating in the creation of the OEKG that contains symbolic event knowledge, including facts, multilingual texts, news articles, images, exemplified questions and their answers. The availability of such trusted resources is of outstanding significance when performing event analytics since the perception of events can heavily depend on the characteristics of the receiver, including political views and cultural influences. While more and more information sources emerge on the Web, specifically on social media, LLMs give nearly unconstrained power of text generation based on undisclosed resources, typically lacking references to amplify further investigations. Consequently, AI-generated misinformation is expected to dominate future disinformation landscapes through deceptive narratives, manipulated images and deepfake videos, making it difficult for both users and algorithms to differentiate between truth and fabrication (Xu et al. 2023). With our research and resources in the CLEOPATRA ITN, we aim to provide a foundation for event analytics that can support the development and evaluation of upcoming technologies for truthful and grounded event analytics.

This book is divided into three parts, focussing on different aspects of event analytics across languages and communities:

Part I: Event-Centric Multilingual and Multimodal NLP Technologies presents recent developments in NLP technologies required to process multilingual information. In particular, this part covers five chapters presenting selected NLP approaches for multilingual information processing that can enhance NLP components to support languages with less resources more effectively, as well as technologies dealing with multimodal information to advance event analytics from heterogeneous sources.

Part II: Event-Centric Multilingual Knowledge Technologies discusses technologies integrating multilingual event-centric information in knowledge graphs and providing user access to such information. The contributions presented in this part include the OEKG, a multilingual event-centric knowledge graph that contains more than 1 million events in 15 languages. Furthermore, this part presents QuoteKG, a knowledge graph of quotes, and methodologies for event recommendation and conversational question answering.

Part III: Event Analytics covers three selected aspects of multilingual event analytics, namely, an analysis of event-centric news spreading barriers, claim detection in social media and the narrativisation of events as a means of presenting event data.

Hannover, Germany Simon Gottschalk

References

Xu D, Fan S, Kankanhalli M (2023) Combating misinformation in the era of generative AI models. In: Proceedings of the 31st ACM International Conference on Multimedia, pp 9291–9298

Zhou C, Li Q, Li C, Yu J, Liu Y, Wang G, Zhang K, Ji C, Yan Q, He L, et al (2023) A comprehensive survey on pretrained foundation models: a history from BERT to ChatGPT. Preprint arXiv:2302.09419

Acknowledgements

The work presented in this book was substantially funded by the European Union's Horizon 2020 research and innovation programme under the Marie Skłodowska-Curie grant agreement no. 812997 (H2020-MSCA-ITN 2018-812997), with the project name CLEOPATRA (Cross-lingual Event-centric Open Analytics Research Academy). Chapters 5 and 12 were written independently, outside the research project.

Special thanks are due to all the research fellows and external partners who contributed to the research and helped to achieve successful results.

Contents

Part I
Event-Centric Multilingual and Multimodal NLP Technologies

Part I of this book consists of five chapters predominantly dealing with research in the novel natural language processing (NLP) approaches applied to texts referring to different types of events. The CLEOPATRA project, as initially proposed, relied on existing and proven methodologies in its NLP activities, covering the usual tasks such as lemmatisation, PoS/MSD tagging, named entity recognition and classification (NERC), dependency parsing, etc. However, the project also started at the very dawn of the introduction of large language models (LLMs) in the NLP processing pipelines that strongly influenced the NLP methodology overall and resulted in a paradigm shift in a couple of years. Consequently, some of the chapters in Part I present research that is to a certain extent still based on previous methods, while other chapters have already adopted LLMs as their methodological core.

Chapter 1 presents a proposition for a language-independent universal named entity recognition (UNER) framework that could be applicable to any language. This proposal is inspired by two similar initiatives in the field of NLP, namely, universal dependencies (UD) and universal tagset (UT). It represents an adapted union of all previous NERC approaches and comes up with a three-level hierarchical classification of named entities that is coupled with the relevant dBpedia entries, thus directly relating names with items in a conceptual data space.

Chapter 2 presents research into the contextualisation of the vast amount of news that is published worldwide with relevant geographic locations. Existing approaches for geolocation estimation were primarily based on either text or photos as separate tasks. Since news photos can lack geographical cues and text can include multiple locations, it is challenging to recognise the focus location of the news story based on only one modality. This chapter introduces novel datasets for multimodal geolocation estimation of news documents, taking into account both text and photos and achieve better results.

Chapter 3 investigates how computational approaches in language typology can improve the results of language classification. Since the CLEOPATRA project was generally oriented towards multilingual processing of events, this chapter offers a novel approach to measuring language distances using several syntactic features from comparable (a corpus of news about EU 2019 elections) or parallel (Parallel

UD) treebanks. This approach can also measure which combination of languages in the training data could improve results in dependency parsing.

Chapter 4 introduces the method developed for sentiment analysis (SA) and detection of hate speech in large multilingual and multimodal news collections. This task is becoming more important every day since the sentiment of a text or a sentence can be crucial for opinion building, while hate speech seems to appear more often than before (although no one has really managed to clearly define what concepts are precisely covered by that term). The research presented here successfully integrates information from multiple modalities to acquire the overall context and then it applies this method to a specific exemplary task.

Chapter 5 concludes Part I with a topic deeply nested in the application of LLMs to the lower-resourced languages. Since, in the most popular multilingual LLMs, the vocabulary of lower-resourced languages is often seriously under-represented in the workpiece dictionaries, this chapter proposes strategies to seek out and protect "vulnerable words" in lower-resourced languages by introducing them into multilingual LLM's dictionaries and providing reasonable initialisations of their embeddings, followed by additional fine-tuning, subject to the limits of available lower-resourced corpora.

The chapters in Part I encompass the set of NLP methods that were developed and then used for multilingual processing in the CLEOPATRA event processing pipeline, demonstrating how language technologies can be successfully coupled with knowledge technologies to achieve better results in automatic processing of event-relevant data.

Marko Tadić

Chapter 1
UNER: Universal Named-Entity Recognition Framework

Diego Alves, Gaurisha Thakkar, and Marko Tadić

Abstract Named-entity recognition and classification (NERC) is an essential natural language processing (NLP) task involved in many applications like interactive question answering, summarising, relation extraction, and text mining. Available NERC corpora follow different annotation schemes that vary in terms of formats and levels of complexity according to research requirements: from 1-level hierarchy annotations (e.g., "Person", "Location", and "Organisation") to multi-level schemes. Inspired by the work of the Universal Dependencies framework in terms of a standard representation of parsed trees, we developed the universal named-entity recognition (UNER) framework, which consists of a multi-level NERC hierarchy and a corresponding workflow that parses data from Wikipedia and DBpedia, translating it to UNER annotations. This chapter presents the UNER hierarchy and its workflow for data extraction and annotation. The proposed process was used to generate an English corpus, which was evaluated qualitatively and quantitatively. Furthermore, seven strategies for annotation improvement were presented and discussed, showing that the usage of information from the Open Event Knowledge Graph (OEKG) can improve our dataset.

1.1 Introduction

Named-entity recognition and classification (NERC) is an essential sub-field of natural language processing (NLP) given the significance of information extraction from texts. It was first defined in 1995 at the 6th Message Understanding Conference (MUC-6) (Chinchor 1998) and has since been utilised in a variety of NLP applications, including events and relations extraction, question answering systems,

D. Alves (✉) · G. Thakkar · M. Tadić
Faculty of Humanities and Social Sciences, University of Zagreb, Zagreb, Croatia
e-mail: dfvalio@ffzg.hr; gthakkar@m.ffzg.hr; marko.tadic@ffzg.hr

I. Marenzi et al. (eds.), *Event Analytics across Languages and Communities*,
https://doi.org/10.1007/978-3-031-64451-1_1

and entity-oriented search. For example, while MUC-7 (Chinchor 1998) defined a 2-level hierarchy with 3–8 nodes per level, the Second Harem NERC schema (Freitas et al. 2010) is composed of 3 levels with 10 to 36 nodes.

To create a universal multilingual annotation named-entity scheme, we introduce the universal named-entity recognition (UNER) framework, a multi-level NERC hierarchy. UNER is based on the Sekine NERC hierarchy (Sekine 2007), with specific changes allowing it to be easily converted to other NERC approaches. Together with this framework, we propose a pipeline for automatically extracting and annotating texts from Wikipedia according to the UNER hierarchy. This pipeline was applied to English Wikipedia, and the resulting UNER corpus was evaluated qualitatively and quantitatively.

In Sect. 1.2, we describe in detail the UNER framework and hierarchy, and in Sect. 1.3, we detail the workflow for data extraction and annotation. Section 1.4 presents the UNER English corpus, as well as the evaluation of this dataset. In Sect. 1.5, we describe the experiments that were conducted to improve the precision and recall of the annotated corpus, and Sect. 1.6 contains our conclusions and potential future research directions.

1.2 UNER Tagging Framework Definition

In this section, we present an overview of the UNER hierarchy and the changes between versions 1 and 2.

1.2.1 UNER: Version-1

The first version of the UNER hierarchy, presented in Alves et al. (2020), is built upon the NERC hierarchy proposed by Sekine (2007), which presents the highest conceptual complexity between the compared NERC schemes (Alves et al. 2020). The number of nodes at each UNER level is described in Table 1.1.

The proposed UNER hierarchy is composed of 4 levels. **Level 0** is the root node, from which all the other levels are derived. **Level 1** consists of three main classes, which correspond to MUC-7 (Chinchor 1998) main categories: *Name*, *Time Expression*, and *Numerical Expression*. **Level 2** is composed of 29 named-entity

Table 1.1 Description of the number of nodes per level inside the UNER hierarchy

Level	Number of nodes
0 (root)	1
1 (classes)	3
2 (categories)	29
3 (types)	95
4 (sub-types)	129

Table 1.2 Description of the number of nodes per level inside the UNER v2 hierarchy

Level	Number of nodes
0 (root)	1
1 (classes)	3
2 (categories)	14
3 (types)	53
4 (sub-types)	88

categories, which are detailed in **Level 3** with 95 types. Additionally, **Level 4** contains 129 sub-types (Alves et al. 2020).

This first version of the UNER hierarchy, therefore, encompasses 215 labels, which can contain up to 4 levels of granularity depending on how detailed is the named-entity type. UNER labels are composed of tags from each level separated by a hyphen, "-". As level 0 is the root and common for all entities, it is not described in the label. For example:

- UNER label *Name-Event-Natural_Phenomenon-Earthquake* is composed of level 1 *Name*, level 2 *Event*, level 3 *Natural Phenomenon*, and level 4 *Earthquake*.

1.2.2 *UNER: Version-2*

The idea of using both Wikipedia data and metadata associated with DBpedia information to generate UNER annotated datasets compelled us to revise the first proposed UNER hierarchy. The main reason is that the automatic annotation process is based on a list of equivalences between UNER labels and DBpedia classes. By generating the list of equivalences, it became apparent that not all UNER labels would have a DBpedia class equivalent. This applies to the vast majority of Time and Numeric expressions. Therefore, we developed version 2 of UNER, presented in the GitHub Web page of the project. It is composed of 124 labels, and its hierarchy is detailed in Table 1.2.

Moreover, in the annotation process, we used the IOB format (Ramshaw and Marcus 1999) as it is widely used by many NERC systems, as shown by Alves et al. (2020). Therefore, each annotated entity token also receives at the beginning of the UNER label the letter "B" if the token is the first of the entity or "I" if inside. Non-entity tokens receive only the tag "O".

1.3 Data Extraction and Annotation

The workflow developed allows the extraction of texts and metadata from Wikipedia (for any language present in this database), followed by the identification of the DBpedia classes via the hyperlinks associated with certain tokens (entities) and

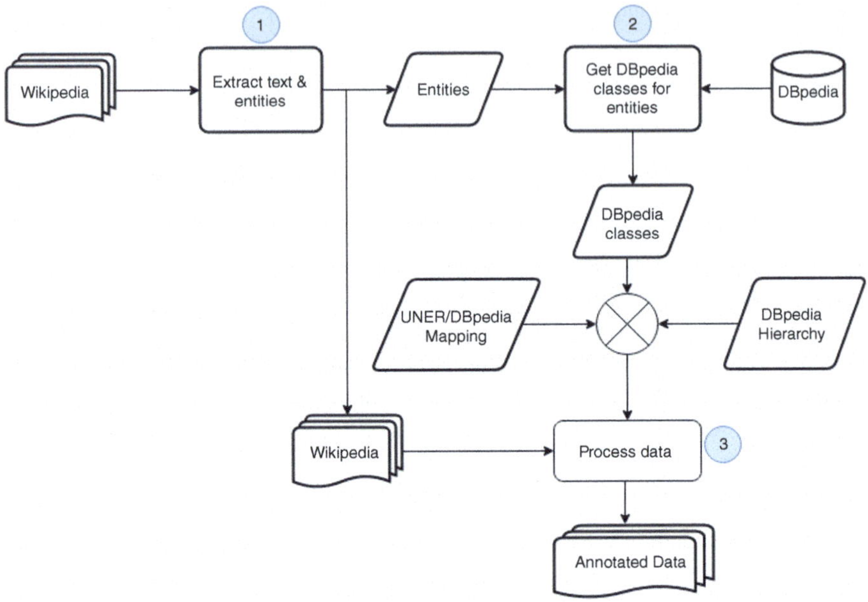

Fig. 1.1 Main processing steps for Wikipedia data extraction and DBpedia/UNER annotations. Squares represent data and diamonds represent processing steps

the translation to UNER types and sub-types (these last two steps being language independent).

Once the main process of data extraction and annotation is over, the workflow proposes post-processing steps to improve the tokenisation, implement the IOB format (Ramshaw and Marcus 1999), and gather statistical information concerning the generated corpus. The whole process is schematised in Fig. 1.1 and is divided into three sub-processes.

1.3.1 Texts and Metadata Extraction

1. **Extraction from Wikipedia dumps**: For a given language, we obtain its latest dump from the Wikimedia Web site.[1] Next, we perform text extraction preserving the hyperlinks in the article using WikiExtractor.[2] These are hyperlinks to other Wikipedia pages, as well as unique identifiers to those named entities. We extract all the unique hyperlinks and sort them alphabetically. We extract the article path

[1] https://dumps.wikimedia.org/

[2] https://github.com/attardi/wikiextractor

from the hyperlinks, eliminating the domain and sub-domain information. The article paths are considered to be named entities.

2. **Wikipedia-DBpedia entity linking**: For all the unique named entities from the dumps, we query the DBpedia endpoint using a SPARQL query with SPARQLWrapper[3] to identify the various classes associated with the entity. This step produces, for each named entity from step 1, a set of DBpedia classes it belongs to.
3. **Wikipedia-DBpedia-UNER back-mapping**: For every extracted named entity obtained in step 1, we use the set of classes produced in step 2, along with a UNER/DBpedia mapping schema, to assign UNER classes to each named entity. For an entity, all the classes obtained from the DBpedia response are mapped to a hierarchy value, the highest valued class is resolved and chosen, and then it is mapped to the UNER class. For constructing the final annotation dataset, we only select sentences that have at least one single named entity. This reduces the sparsity of annotations and thus reduces the false-negative rate in our test models. This step produces an initial tagged corpus from the whole Wikipedia dump for a specific language.

1.3.2 Tagging Process

1. **UNER/DBpedia Mapping**: This mapper links every DBpedia class to one UNER tag. Extracted named entities may have multiple DBpedia classes. It assigns the best-suited UNER tag to each DBpedia class. For example, entity ***2015 European Games*** has the following DBpedia classes with the respective UNER equivalences:

   - **dbo:Event**—*Name-Event-Historical-Event*
   - **dbo:SoccerTournament**—*Name-Event-Occasion-Game*
   - **dbo:SocietalEvent**—*Name-Event-Historical-Event*
   - **dbo:SportsEvent**— *Name-Event-Occasion-Game*
   - **owl:Thing**—*NULL*

   The value on the left represents a DBpedia class, while its UNER equivalent appears on the class's right. It maps all DBpedia classes to their UNER counterparts.
2. **DBpedia Hierarchy**: This mapper assigns each DBpedia class a priority. This is used to select a specific DBpedia class from the associated collection of classes. The following are examples of classes and their priorities:

   - **dbo:Event**—2
   - **dbo:SoccerTournament**—4

[3] https://rdflib.dev/sparqlwrapper/

- **dbo:SocietalEvent**—2
- **dbo:SportsEvent**—4
- **owl:Thing**—1

For entity ***2015 European Games***, the DBpedia class **SoccerTournament** presides over the other classes, as it has a higher priority value. If the extracted entity has two assigned classes with the same hierarchy value, the first from the list is chosen as the final one. All the DBpedia classes were assigned with a hierarchy value according to the DBpedia Ontology,[4] where classes are presented in a structural order, which allowed us to define the hierarchical levels.

1.3.3 Post-processing Steps

The post-processing steps correspond to three different scripts that provide:

1. The improvement of the tokenisation (using regular expressions) by isolating punctuation characters that were connected with words. In addition, it applies the IOB format (Ramshaw and Marcus 1999) to the UNER annotations inside the text.
2. The calculation of the following statistical information concerning the generated corpus: Total number of tokens, Number of Non-entity Tokens (tag "O"), Number of Entity Tokens (tags "B" or "I"), and Number of Entities (tag "B"). The script also provides a list of all UNER tags with the number of occurrences of each tag inside the corpus.
3. Listing the entities inside the corpus (tokens and the corresponding UNER tag). Each identified entity appears once in this list, even if it has multiple occurrences in the corpus.

The whole process and post-processing steps were applied to the English language, generating the UNER English corpus, which is described and evaluated in the following section. This baseline corpus is the base for the improvement experiments presented in later sections.

1.4 UNER English Corpus (Baseline)

In this section, we present in detail the UNER English corpus together with the evaluation campaign, which was conducted to check the overall quality of the data.

[4] https://github.com/cleopatra-itn/MIDAS/blob/master/uner-documentation/DbOC.md

Table 1.3 Corpora annotation statistics

	English UNER corpus
Total number of tokens	325, 395,838
Number of non-entity tokens	320, 719,350
Number of entity tokens	31, 676,488
Number of entities	15, 101,318
Number of different entities	630,519

Table 1.4 Corpora annotation statistics in terms of the number of occurrences of the most used NERC classes (and % of all entities occurrences)

	English UNER corpus
Person	4,200,313 (27.8%)
Location	2,613,248 (17.3%)
Organisation	3,489,813 (23.1%)

1.4.1 General Information

The English Wikipedia[5] is composed of 6,188,204 articles (3.3 GB). After applying the main process of the proposed workflow, we obtained annotated text files divided into folders (17,150 files in 172 folders) (Alves et al. 2021).

Statistical information concerning the corpus is obtained by applying the post-processing steps previously described. Table 1.3 presents the main statistics about the number of tokens and entities. Inside the UNER English corpus, 8.9% of tokens are entities. And, in Table 1.4, we present the statistics concerning the most frequent NERC classes.

As presented in Sect. 1.2.2, the UNER hierarchy used for annotating the English Wikipedia texts is composed of 124 different multi-levelled labels with equivalences to DBpedia classes. However, in the UNER English corpus, only 99 different UNER tags (80% of the total) occurred.

As explained previously, the UNER hierarchy is composed of categories, types, and sub-types. The UNER includes the most common classes used in NERC (*Person*, *Location*, *Organisation*) in its hierarchical Level 2. Therefore, it is possible to analyse the generated corpora in terms of these widespread generic classes.

These three classes cover 68.2% of named entities in the generated corpus.

1.4.2 Qualitative Evaluation

An analysis of 943 entities randomly selected from the UNER English Corpus has been performed to evaluate this step of the workflow. For each one, we have checked

[5] http://en.wikipedia.org/w/index.php?title=English%20Wikipedia&oldid=987449701

Table 1.5 Evaluation of the annotation step: DBpedia class extraction and translation to the UNER hierarchy

Tag evaluation	Number of occurrences	Percentage
Correct	797	85%
Correct but vague	55	6%
Incorrect due to DBpedia	62	7%
Incorrect due to UNER association	29	3%

the DBpedia-associated classes and the final UNER chosen tag. Table 1.5 presents the results of this evaluation.

In the selected sample, 91% of the entities are correctly tagged with UNER tags. Nevertheless, 6% are associated with the correct UNER type but with a generic subtype. For example, ***Bengkulu*** should be tagged as *Name-Location-GPE-City* but received the tag *Name-Location-GPE-GPE_Other*. Errors may come from mistakes in the DBpedia classes associated with the tokens or due to the prioritisation rules and equivalences defined between DBpedia and UNER.

1.4.3 UNER English Golden Dataset

Besides the statistical information presented above, a sample from the generated corpus has been selected and corrected using WebAnno (Eckart de Castilho et al. 2016) by one annotator. The sample corresponds to one entire file from the output folder and contains 519 sentences and 105 different UNER labels. The annotations were done by a non-native English speaker who is a member of the project, following objective guidelines. In cases of multi-possible assignments, a final choice was made by the annotator so that each entity would have only one label in the golden set. Table 1.6 presents the evaluation results of the baseline annotations of the file used to create the golden dataset in terms of Precision, Recall, and F1-measure, considering the mean value of all 105 labels for each metric.

As explained previously, the annotation of a particular named entity depends on the existence of hyperlinks. However, these links are not always associated with the tokens if the entity is mentioned repeatedly in the article. This may be one of the main reasons for the low value obtained for recall.

Table 1.6 Precision, recall, and F1-measure values of UNER EN dataset considering 519 manually annotated sentences

Experiment	Precision	Recall	F1-measure
Baseline	61.9	27.2	37.8

1.5 Dataset Improvement

Evaluation of the baseline annotated file using the golden UNER English corpus shows that the automatic annotation workflow has room for improvement, especially in terms of reducing the number of false negatives. Strategies for completing the annotation using dictionaries and knowledge graphs were applied to the English corpus. The ensemble of experiments and the evaluation is presented below.

1.5.1 Experiment Design

Seven different experiments were conducted:

1. Global Dictionary: From the whole UNER English corpus, we established a dictionary of single-word entities and the respective UNER label. As the same entity may appear in the corpus with different UNER tags (due to the associated DBpedia classes), we selected for each entity the label with the highest number of occurrences. This dictionary is then used to complete the annotations of the corpus. Only entities with lengths longer than two characters were considered, and numerical entities were excluded from the dictionary. The final size of the global dictionary is 826,371 entities.
2. Global Dictionary only with multi-word entities: Similar to the previous experiment; in this case, only entities with more than one token were considered. In total, the global dictionary is composed of 665,081 multi-word entities.
3. Local Dictionaries: In this setup, we processed every Wikipedia dump file as a single article and applied the strategy "one meaning per discourse". Every entity in the article that is linked to the UNER is cached into a local lookup dictionary, with its text as the key and UNER class as the value. For every subsequent occurrence of the key in the given article, we annotated the text with the corresponding UNER class. We performed this step with the speculation that entities are more likely to appear within a single article than in a completely unrelated article. For example, *Barack Obama* as a person is more likely to appear in an article describing him as president than as a fictional character who appears in fictional content about him.
4. Global OEKG Dictionary: Open Event Knowledge Graph (OEKG)[6] is a multilingual event-centric resource. Its instances have specific DBpedia classes; therefore, we intersected all the single-word entries from the global dictionary with elements from the OEKG. For each entity, its associated DBpedia class from the OEKG was then mapped to the UNER. The global OEKG dictionary contains 128,813 entries.

[6] http://cleopatra-project.eu/index.php/open-event-knowledge-graph/

5. Global OEKG Dictionary only with multi-word entities: Similar to experiment 4; only in this case, only entities with more than one token were considered (110,226 entities in total).
6. Local Dictionaries followed by Global OEKG Dictionary: Combination of experiment 3 with completion of annotations using dictionary established for experiment 4.
7. Local Dictionaries followed by OEKG Dictionary only with multi-word entities: Corpus from experiment 3 is completed using the dictionary from experiment 5.

In all experiments, dictionaries were ordered from the longest entities to the shortest ones ("longest match" strategy) to guarantee that preferably multi-word entities were annotated and not single-word ones.

1.5.2 Evaluation

The evaluation was conducted using the golden corpus presented previously. The baseline is the correspondent file with automatic annotations as a result of the workflow described in Sect. 1.4.

The golden corpus has 105 different UNER labels; however, the baseline annotated file has only 62. For each possible label, we calculated precision, recall, and F1-measure. The IOB format (Ramshaw and Marcus 1999) was applied; therefore, each UNER label can start either with "B" or "I", and non-entity tokens were tagged with "O".

From the 62 labels of the baseline, only 45 presented results different from 0. Therefore, the values present in the following Table 1.7 consider only these tags and represent the mean value of all the tags taken into account. Table 1.7 presents the metrics obtained for the baseline and each one of the experiments described in the previous subsection.

The global dictionary approach (experiment 1) provides the highest value of recall (+3.7 compared to the baseline), but precision is considerably lower (−40.8). A similar situation is observed when the global dictionary is used only with multi-token entities (experiment 2). Other experiments do not decrease precision so

Table 1.7 Precision, recall, and F1-measure values of experiments for improving the UNER EN dataset

Experiment	Precision	Recall	F1 measure
Baseline	72.9	32.0	39.2
1	32.1	**35.7**	27.0
2	47.5	34.8	34.0
3	73.6	29.6	36.8
4	71.1	33.9	**40.8**
5	73.0	33.4	40.5
6	72.1	32.1	39.1
7	**74.0**	31.5	38.7

Table 1.8 Precision, Recall, and F1-measure values of experiments for improving UNER EN dataset considering only level 3 of the UNER hierarchy

Experiment	Precision	Recall	F1 measure
Baseline	**76.9**	25.1	34.0
1	25.9	**31.0**	25.2
2	37.2	27.8	31.1
3	76.8	24.5	33.2
4	74.6	27.3	**36.0**
5	76.6	26.3	35.3
6	74.6	26.9	35.5
7	76.6	25.8	34.7

drastically, and in some cases, this metric is even increased. The recall is increased compared to the baseline for all experiments except for 3 and 6, 7. The use of local dictionaries was not an effective solution for improving this evaluation metric.

The best option, considering the F1 measure, is the usage of the dictionary verified with the OEKG (experiment 4). Precision is slightly lower than the baseline (−1.8), while recall and F1 measures are higher (+1.9 and +1.6, respectively).

The evaluation of each experiment considering only this upper level of the UNER hierarchy is presented in Table 1.8. The IOB format was also considered; therefore, UNER labels could be preceded by either "B" or "I", and non-entity tokens were tagged with "O".

In this scenario, the highest precision is from the baseline. The best recall is obtained when the global dictionary is used (experiment 1), but as was observed before, in this case, precision is heavily impacted compared to the baseline (-51.0). Experiment 4 is the one with the highest F1 measure, the same as the previous evaluation, where all UNER levels were considered.

Thus, concerning the improvement experiments, the best-identified option was to use a dictionary fine-tuned from the Open Event Knowledge Graph. This resource allows more precise identification of the specific DBpedia classes and therefore helps improve recall without considerable loss in precision.

1.6 Conclusions and Future Directions

In this chapter, we described the UNER hierarchy, which is intended to serve as a universal framework for NERC. Moreover, we described an automatic workflow for generating multilingual named-entity recognition corpora by using Wikipedia and DBpedia data and following the UNER hierarchy. The whole process is available as an open source and can be applied to any language having Wikipedia and DBpedia.[7]

We also presented the English UNER corpus that has been generated using the proposed pipeline. This dataset has been described and evaluated with a manually

[7] https://github.com/cleopatra-itn/MIDAS

annotated golden set. While the precision score was higher than 60, the recall value was lower than 30. Thus, an ensemble of experiments was conducted to improve the final annotated dataset.

We have identified that the best results were obtained by using a dictionary of entities with verification of the associated DBpedia class using the Open Event Knowledge Graph: 76.9 for precision, 31.0 for recall, and 36.0 for F1 measure. Nevertheless, these results show that there is still room for improvement in both recall and F1 measure.

As perspectives for future work, our main focus is the improvement of the recall to obtain a more efficient pipeline, which can, then, be used to generate UNER corpora for all languages available on Wikipedia. With the generated corpora, deep-learning models can be trained for automatic NERC. Furthermore, the hierarchy should also be completed with more elaborated temporal labels, which were excluded in UNER v.2.

Acknowledgments The work presented in this chapter has received funding from the European Union's Horizon 2020 research and innovation program under the Marie Skłodowska-Curie grant agreement no. 812997 and under the name CLEOPATRA (Cross-lingual Event-centric Open Analytics Research Academy).

References

Alves D, Kuculo T, Amaral G, Thakkar G, Tadić M (2020) UNER: Universal Named-Entity Recognition Framework. In: Proceedings of the CLEOPATRA Workshop at the 19th International Semantic Conference, CEUR Workshop Proceedings

Alves D, Thakkar G, Tadić M (2021) Building and evaluating universal named-entity recognition english corpus. In: Proceedings of the 2nd International Workshop on Cross-lingual Event-centric Open Analytics, CEUR Workshop Proceedings, pp 2–16

Eckart de Castilho R, Mújdricza-Maydt É, Yimam SM, Hartmann S, Gurevych I, Frank A, Biemann C (2016) A web-based tool for the integrated annotation of semantic and syntactic structures. In: Proceedings of the Workshop on Language Technology Resources and Tools for Digital Humanities (LT4DH), The COLING 2016 Organizing Committee, Osaka, pp 76–84. https://aclanthology.org/W16-4011

Chinchor N (1998) Appendix E: MUC-7 named entity task definition (version 3.5). In: Seventh Message Understanding Conference: Proceedings of a Conference Held in Fairfax, Virginia, MUC 1998, April 29–May 1, 1998. ACL. https://aclanthology.org/M98-1028/

Freitas C, Carvalho P, Gonçalo Oliveira H, Mota C, Santos D (2010) Second HAREM: advancing the state of the art of named entity recognition in Portuguese. In: Proceedings of the International Conference on Language Resources and Evaluation (LREC 2010). European Language Resources Association, pp 3630–3637

Ramshaw LA, Marcus MP (1999) Text chunking using transformation-based learning. In: Natural language processing using very large corpora. Springer, Berlin, pp 157–176

Sekine S (2007) The Definition of Sekine's Extended Named Entities. https://nlp.cs.nyu.edu/ene/version7_1_0Beng.html. Accessed 28 Feb 2020

Chapter 2
Multimodal Geolocation Estimation in News Documents

Golsa Tahmasebzadeh, Eric Müller-Budack, and Ralph Ewerth

Abstract With the proliferation of news documents on the Internet, online news reading has become an important approach for information acquisition in people's daily lives. There has, however, been increasing concern with the growing infusion of misinformation. As a complement to news text, associated photos provide readers with additional information to facilitate their ability to find the information they need. To contextualise the vast amount of news that is published worldwide, the geographic content is crucial. On the other hand, the geographic content plays an important role in news recommendation to facilitate user desires. Existing approaches for geolocation estimation are primarily based on either text or photos as separate tasks. However, news photos can lack geographical cues, and text can include multiple locations. Therefore, it is challenging to recognise the focus location of the news story based on only one modality. We introduce novel datasets for multimodal geolocation estimation of news documents. We evaluate current methods on the benchmark datasets and suggest new methods for news geolocalisation using textual and visual content. In addition, we introduce a news retrieval system called GeoWINE based on the geographic content of news photos to emphasise the importance of geolocation estimation in the news domain.

G. Tahmasebzadeh (✉)
L3S Research Center, Leibniz Universität Hannover, Hannover, Germany
e-mail: golsa.tahmasebzadeh@tib.eu

E. Müller-Budack
TIB – Leibniz Information Centre for Science and Technology, Hannover, Germany
e-mail: eric.mueller@tib.eu

R. Ewerth
TIB – Leibniz Information Centre for Science and Technology, Hannover, Germany

L3S Research Center, Leibniz Universität Hannover, Hannover, Germany
e-mail: ralph.ewerth@tib.eu

I. Marenzi et al. (eds.), *Event Analytics across Languages and Communities*,
https://doi.org/10.1007/978-3-031-64451-1_2

2.1 Introduction

Every single day, new events arise in different parts of the world, and social media and the Web allow the dissemination of news in diverse modalities such as image and text and in different languages. Therefore, it is important to find ways to manage the flow of information, to consume news from a variety of sources, and to maintain a balanced perspective. One of the key components of an event is the location that it refers to. Since news documents are usually coupled with photos, finding the location where the photo was taken is an important aspect for various real-world applications. Example applications are news retrieval (Armitage et al. 2020), image verification (Cheng et al. 2019), and misinformation detection in news (Singhal et al. 2019), to name a few. Most prior methods for predicting the geolocation of photos rely exclusively on visual data (Izbicki et al. 2019; Kim et al. 2017; Müller-Budack et al. 2018), and only a handful of techniques utilise multiple modalities (Kordopatis-Zilos et al. 2017, 2016). Existing image-based methods are mainly focused on specific environments such as cities (Berton et al. 2022; Kim et al. 2017) or landmarks (Avrithis et al. 2010; Boiarov and Tyantov 2019; Weyand et al. 2020).

The majority of multimodal techniques utilise the Yahoo Flickr Creative Commons 100 Million (YFCC100M) dataset (Thomee et al. 2015) and rely on the tags associated with the images. However, they fail to leverage the detailed textual information contained in news articles that could suggest potential locations of photos. (Fig. 2.1b[1]). The *BreakingNews* dataset (Ramisa et al. 2018) is a multimodal collection of news articles that includes geolocation labels. These labels are primarily obtained from the Resource Description Framework (RDF) Site Summary (RSS) or, when unavailable, deduced through methods like analysing the location of the publisher or the text of the story. However, the geolocations obtained may

Fig. 2.1 Samples from the *MMG-NewsPhoto* dataset. GT: Ground Truth location. Photos are replaced with similar ones due to licence restrictions

[1] The photos are licenced as follows: (a) by Terra Eclipse and (b) by Aron Urb: https://creativecommons.org/licenses/by/2.0/.

be imprecise or incorrect. Additionally, a limitation of the *BreakingNews* dataset is that the test split labels are generated through the same process. Overall, there is a significant need for multimodal news article datasets that provide geolocation labels specifically for news photos, alongside multimodal approaches for accurately determining the geolocations of in news documents.

In this chapter, we define the task of geolocation estimation as a multimodal problem. We suggest multimodal approaches that integrate both visual and textual information from the news photo and the body text to automatically identify the focus location of the whole news story (Tahmasebzadeh et al. 2022) or the geolocation of the photo (Tahmasebzadeh et al. 2023). The main contributions are summarised as follows: (1) We introduce two datasets for the task of geolocation estimation in news. The *MMG-NewsPhoto* (Multimodal Geolocation Estimation of News Photos) contains more than half a million news articles. The proposed dataset covers more than 14,000 cities and 241 countries across all continents within multiple news domains such as *Health*, *Business*, *Society*, and *Politics* and *MM-Locate-News* (Multimodal Focus Location Estimation in News), which consists of 6.395 news articles covering 237 cities and 152 countries across all continents as well as multiple domains. (2) We propose detailed annotation instructions and identify visual concepts specific to news that indicate the geographic location of a photo. (3) We introduce multimodal methods that utilise state-of-the-art visual and textual features for the geolocalisation of news documents. (4) Our methods are assessed on the datasets we introduced, and we compare their performance with state-of-the-art techniques, as well as some baseline re-implementations. (5) To highlight the significance of geolocation estimation in analysing news content, we introduce our proposed systems that benefit from the geolocation of a photo as a core task for information retrieval (Tahmasebzadeh et al. 2020) or photo verification (Tahmasebzadeh et al. 2021).

The remainder of the chapter is structured as follows: Section 2.2 describes the related work. In Sect. 2.3, we introduce our proposed datasets. The proposed models for multimodal geolocation estimation are presented in Sects. 2.4 and 2.5, while the methods for information retrieval are discussed in Sect. 2.6. Section 2.7 concludes the chapter and outlines limitations and potential future directions.

2.2 Related Work

There are two main criteria to classify the approaches for geolocation estimation of photos: the environment target and the data type, i.e. images and multimodal data (Brejcha and Cadík 2017). In this section, we briefly review related work on photo geolocation estimation and primarily focus on multimodal approaches, existing datasets, and their drawbacks.

Image-Based Approaches Many existing methods based on image geolocalisation focus on urban (Berton et al. 2022; Kim et al. 2017) and natural environments, such as mountains (Baatz et al. 2012; Tomesek et al. 2022). Some attempts estimate photo location at global scale without any prior assumptions about the environment. Most of them treat geolocation estimation as a classification problem (Müller-Budack et al. 2018; Seo et al. 2018; Theiner et al. 2022; Weyand et al. 2016). Improvements were made, for example, by exploiting a retrieval approach and a large geo-tagged image database (Vo et al. 2017), using overlapping sets of visually similar cells (Seo et al. 2018), incorporating a hierarchical cell structure as well as environmental scene context (Müller-Budack et al. 2018), or leveraging the advantages of contrastive learning (Kordopatis-Zilos et al. 2021). However, while these approaches achieve promising results solely based on visual information, news provides textual information that can further increase the performance, particularly in the absence of distinct geographical cues (Fig. 2.1b).

Multimodal Approaches There are only a few methods (Crandall et al. 2009; Kordopatis-Zilos et al. 2017, 2016; Ramisa et al. 2018; Serdyukov et al. 2009) that address geolocation estimation as a multimodal problem, most of which rely on constructing large-scale geographical language models by generating a probabilistic model based on mentions of textual tags across the globe (Kordopatis-Zilos et al. 2017, 2016; Serdyukov et al. 2009). Crandall et al. (2009) combine image content and textual metadata at two levels of granularity: at city level ($\approx$100 km) and landmark level ($\approx$100 m). Trevisiol et al. (2013) process the textual information of a set of videos to determine their geo-relevance and to find frequent matching items. Where such information is not present, they resort to visual features. Later, a multimodal approach was proposed by Ramisa et al. (2018), where they combine visual features with text using the nearest neighbour method and support vector regression (SVR).

Multimodal Datasets Most multimodal approaches are based on the *YFCC 100M* dataset (Thomee et al. 2015) or the *MediaEval Placing Task* benchmark datasets Larson et al. (2017) including images, videos, and metadata. Another dataset proposed by Uzkent et al. (2019) contains images and text from Wikipedia combined with satellite images. More recently, a dataset called *Multiple Languages and Modalities* (MLM) (Armitage et al. 2020) has been introduced, which includes images along with multilingual texts from *Wikidata* (Vrandecic and Krötzsch 2014). Unlike the previous datasets, the *BreakingNews* introduced by Ramisa et al. (2018) contains multimodal news articles and is the most relevant for our work. It includes image, text, caption, and metadata (such as geo-coordinates and popularity) and covers various domains such as *Sports*, *Politics*, and *Health*. The provided geolocation labels for both training and evaluation are extracted from the RSS, publisher, or news text. But as discussed in Sect. 2.1, these automatically derived locations can be inaccurate or even wrong. Instead, we provide high-quality manually annotated photo geolocations for fair and reliable evaluation (Tahmasebzadeh et al. 2022, 2023).

2.3 Proposed Datasets

This section provides an overview of the proposed datasets for the task of multimodal geolocation estimation: *MMG-NewsPhoto* (Tahmasebzadeh et al. 2023) and *MM-locate-news* (Tahmasebzadeh et al. 2022). Both datasets are comprised of image-text pairs of news documents labelled with geolocations. However, there is a slight difference in what exactly the labels mean. In the MMG-NewsPhoto dataset, the labels only represent the geolocation of the photo. However in *MM-locate-News*, the labels represent not only the location where the photo was taken but also the focus location of the news body text. In the following sections, we discuss the details of both datasets.

2.3.1 MMG-NewsPhoto *Dataset*

In this section, we explain the dataset creation and annotation process of the proposed *MMG-NewsPhoto* dataset for multimodal geolocation estimation of news photos.

2.3.1.1 Dataset Creation

We use the collection of articles provided by the *Good News* (Biten et al. 2019) and *CC-News* (Mackenzie et al. 2020) datasets. *Good News* (Biten et al. 2019) is an image-captioning dataset comprising 466,000 image-caption pairs. Based on Web links to the news articles, we extract all articles with a body text, title, image link(s) with corresponding caption(s), and domain label(s). *CC-News* (Mackenzie et al. 2020) includes 44 million documents written in English extracted from around 30,000 unique news sources. We sort the sources based on the number of news articles and scrape news documents from the top-20 sources in the same way mentioned above. Finally, we download all the images and discard the ones with corrupted or inaccessible images. As a result, we end up with circa 10 million data samples, including body text, and at least one image caption pair per sample acquired from both news sources.

Initial Removal We remove redundant documents (except one) based on the cosine similarity (normalised to [0, 1]) of the body texts using *TF-IDF* (Term Frequency; Inverse Document Frequency) above a threshold of 0.5. Next, we manually group the domain labels into ten categories such as *Health*, *Business*, and *Politics* (see full list in Fig. 2.2, left). Some domains such as *Art* and *Technology* include various invalid images for the task, i.e. ads or stock photos. We discard these types of images as they typically lack geographic content or do not correspond to the locations mentioned in the body text of news.

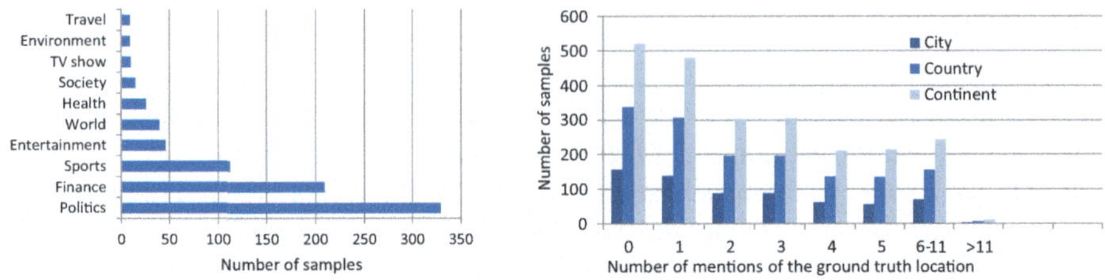

Fig. 2.2 Left: Test data distribution among domains. Right: Frequency of ground-truth location mentions in the body text for the test split

Location Linking We assume that locations mentioned in a caption are good candidates for photo geolocation. We apply named entity recognition and disambiguation to extract all locations in the captions. Following related work (Müller-Budack et al. 2021), we use *spaCy* (Honnibal et al. 2020) to extract the named entities and use *Wikifier* (Brank et al. 2018) to link them to *Wikidata* entities. We only keep entities of type *Location* with valid geocoordinates (latitude, longitude) extracted from the *Wikidata* Property *P625*.

Photo Location Assignment The location entities extracted from the captions do not always indicate the photo locations and can, for example, also refer to entity attributes, e.g. "*US President Biden*". Thus, captions are tokenised to extract certain prepositions, e.g. "across", "along", and "in", which, combined with a location mention, are more likely to refer to the photo location. We keep samples for which the distance of 1 of 37 prepositions[5] to the *claimed photo location* is at most two tokens. Furthermore, samples with more than one unique location are removed, resulting in exactly one *claimed photo location*.

Location Enrichment We apply reverse geocoding to map around 50,000 fine-grained locations (i.e. city, road, building, etc.) extracted from the captions to cities using Nominatim.[2] Next, we extract associated country (*Property P17*), continent (*Property P30*), and geo-coordinates (*Property P625*) from *Wikidata*.

Data Sampling For manual annotation, 3.000 samples are selected to construct the test dataset. To avoid bias, the samples are selected (1) from all domains, (2) from all continents, (3) from highly populated cities (minimum population of 500,000) and medium populated cities (population, 20,000–500,000), (4) with at least three unique locations mentioned in text, and (5) with a different number of mentions of the ground-truth location in the body text. The latter ensures that simple cases with frequent mentions of the ground truth and complex cases, i.e. many locations mentioned in the text with somewhat equal frequencies, are included. For simple cases, a textual approach that leverages the frequency of named entities can already achieve high performance without even considering the image. Based on complex

[2] https://nominatim.org/release-docs/latest/api/Reverse/

cases, we can analyse the direct impact of the image for multimodal geolocation estimation. The statistics for the test split are visualised in Fig. 2.2, right. From the remaining, 10% are randomly chosen for validation, and the rest is used for training.

2.3.1.2 Data Annotation Process

We give an in-depth explanation of the guidelines used for the manual annotation of the test split, which is aimed at making the assessment fair and transparent. The exact guidelines used during annotation are provided on our *GitHub* page[5].

Geo-representative Concepts For photo geolocation estimation, a *geographically representative image* depicts concepts that help identify its location. We group *geo-representative concepts* into two types: *strong* and *weak concepts*. A *strong concept* is a unique identity of a location, e.g. the appearance of the *Eiffel Tower* in an image that can unambiguously be assigned to the city *Paris*, country *France*, and continent *Europe*. A *weak concept*, on the other hand, provides clues for one or even a few specific locations but without sufficient evidence on its own. For example, a certain *President* is an identity of a country but can travel to different locations. Only multiple *weak concepts*, all of which correspond to the same location, in an image can lead to the identification of the geolocation of news photos. For instance, multiple *car plates* or *groups of people* can represent the corresponding country. As shown in Table 2.1, we define *strong* or *weak* visual concepts based on the following eight categories: *building*, *clothing*, *event*, *group of people*, *natural scenery*, *object*, *public personality*, and *scene text*.

Annotation Questions (Q). Given an image-caption pair and the linked location of the caption, we ask each annotator the following questions: **Q1:** ***Is it a valid sample?*** To determine whether a sample is valid for the identification of the photo geolocation, an annotator selects *"no"* if (1) the image is an advert, a stock photo, a Web page, a map, or a data visualisation and (2) the linked location is wrong, not a location, or not the *claimed photo location* (see paragraph *Photo Location Assignment*) of the caption. Otherwise, *"yes"* is chosen. **Q2:** ***Which weak and strong concepts are shown in the image?*** The annotator selects the strong or weak concepts (Table 2.1) depicted in the image. **Q3:** ***Is the linked city (Q3.1), country (Q3.2), continent (Q3.3) shown in the image?*** These questions are asked to obtain the ground-truth location at various granularities. A user selects *"yes"* if (1) at least one *strong concept* is visible, (2) a single *weak concept* occurs in high frequency (e.g. multiple *car plates*), (3) a combination of at least two distinct *weak concepts* is shown, or (4) a single *weak concept* with valid proof (e.g. a Web page that proves the location) is provided. Otherwise, *"no"* is selected. If *"yes"* is given as an answer, a confidence level is selected: *"very confident"*, *"confident"*, and *"not confident"*. **Q4:** ***What is the environmental setting of the image?*** The user selects one of the following categories, *"indoor"*, *"outdoor urban"*, *or "outdoor nature"*, to indicate the environment in which an image was taken. **Q5:** ***Is it a closeup?*** Since locations are usually difficult to predict for closeups, we asked

Table 2.1 Strong and weak visual concepts used in the annotation process

Strong geo-representative concepts			
Category	City	Country	Continent
Building	Buildings, landmarks	–	–
Clothing	–	Public service uniforms	–
Event	Social movements, sports competitions	Social movements, sports competitions, natural disasters, country elections, wars	Sports competitions, natural disasters
Group of people	-	-	-
Natural scenery	City-specific natural landmarks	Country-specific natural landmarks	Continent-specific natural landmarks
Object	Logos of events, organisations, etc.	Public service vehicles	–
Public personality	–	–	–
Scene text	Street signs with mentions of cities	Country names in signs	–
Weak geo-representative concepts			
Category	City	Country	Continent
Building	–	Buildings with specific architectures	–
Clothing	Uniforms of sport clubs	Uniforms of soldiers, cultural costumes, national sport team uniforms	–
Event	–	–	–
Group of people	–	Residents of a country, common activity	–
Natural scenery	–	–	Land forms, flora, fauna
Object	–	Personal cars and/or car plates, flag, logo	–
Public personality	–	Politicians, athletes, celebrities	–
Scene text	–	Text in specific language	–

the annotators to identify whether the image shows a closeup or not. **Q6:** ***Did you need external resources for Q3?*** The final question determines whether or not the annotator needed external resources to decide on Q3. If "Yes" is selected, we asked the annotators to provide the links.

Annotator Training We employed four graduate students with computer science backgrounds who were paid 10 EUR per hour (slightly above the minimum wage in Germany in early 2022) for annotations. Furthermore, three experts (doctoral and postdoctoral researchers) with a research focus on computer vision and multimodal analytics provided annotations. All annotators were trained based on the annotation guidelines[5]. We performed two dry runs using 100 samples and discussed the results to refine the guidelines. **Annotation Process.** The annotation task was performed in two steps as follows: (1) All annotators were asked to validate the 3,000 samples according to Q1. Using majority voting, 1,700 valid samples were obtained. (2) For each *valid* sample, Q2 to Q6 were annotated by three annotators, and majority voting was applied to select samples where two users agreed on the answer per question.

Based on selected answers for Q3.1 to Q3.3, we obtained the final annotations. For all questions, the answer should be *"yes"*, with a confidence level of either *"very confident"* or *"confident"*. Samples where at least two annotators selected the confidence level *"not confident"* were re-annotated by an expert. As a result, we obtained final annotations for Q3.1, Q3.2, and Q3.3, where the answers correspond to the granularity of the geolocation of images. These granularities are turned into three variants of the test data: Test_{city}, $\text{Test}_{country}$, and $\text{Test}_{continent}$. Please note that finer granularity samples are subsets of coarser granularities.

Annotation Study Findings Krippendorff's alpha (Krippendorff 2011) was used to calculate inter-annotator agreements for Q3. The agreements are 0.41 for *city*, 0.41 for *country*, and 0.51 for the *continent*, which we consider low to moderate. Responses to Q4 and Q5 indicated that 40.2% of the images are closeups and 37.7% are indoor images, both of which typically depict few weak geo-representative concepts and are challenging for the photo-geolocation task. For 49.7% of the samples, annotators needed external resources (Q6) to decide whether the image showed the linked location. Overall, these numbers demonstrate the difficulty of the task for humans and explain the moderate inter-coder agreement for Q3.

Dataset Statistics The *MMG-NewsPhoto* contains 554,768 training, 60,893 validation, and 2,259 test samples (sum for all granularities). The dataset contains 14,331 cities, 241 countries and 6 continents. Table 2.2 shows data distribution among continents and top ten countries. Since 1,700 test samples and thus about 57% of the test samples are valid, we assume that training and validation sets contain a similar proportion of valid samples.

Table 2.2 Data distribution for continents (top) and top 10 countries (bottom)

				Europe	N. America	Asia	Oceania	Africa	S. America	Total
Train				190,064	188,175	121,045	20,468	21,096	13,920	554,768
Validation				21,041	20,675	13,120	2,147	2,331	1,579	60,893
Test_{city}				196	189	215	13	27	20	660
$\text{Test}_{country}$				235	212	274	13	35	25	794
$\text{Test}_{continent}$				235	215	278	13	37	27	805
Total				211,769	209,466	134,932	22,654	22,526	15,573	617,920
	U.S.	U.K.	India	China	Australia	France	Japan	Germany	Spain	Russia
Train	173,584	82,917	27,435	18,390	17,018	16,347	15,669	14,477	13,702	9,330
Validation	19,076	9,253	3,024	2,007	1,805	1,766	1,732	1,569	1,459	1,055
$\text{Test}_{country}$	190	82	121	11	11	8	17	24	11	15

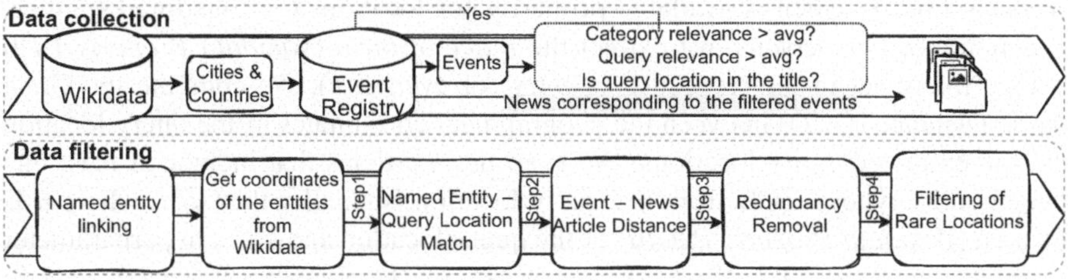

Fig. 2.3 MM-Locate-News data collection and filtering steps

2.3.2 MM-locate-News *Dataset*

This section presents a novel dataset called *Multimodal Focus Location of News (MM-locate-News)*.[3] In the sequel, data collection and cleaning steps (Fig. 2.3) as well as annotation process and dataset statistics are presented.

2.3.2.1 Dataset Creation

Data Collection The dataset has been collected in a weakly supervised fashion. To cover a variety of locations from all six continents, we extract all countries, capitals, highly populated cities, and medium-populated cities from Wikidata (Vrandecic and Krötzsch 2014). For each location, we query *EventRegistry*[4] for events between 2016 and 2020 from the following categories: *sports*, *business*, *environment*, *society*, *health*, and *politics*. Note that *EventRegistry* automatically clusters news articles reporting on the same (or similar) events and that the news title of the cluster centroid represents the event name. To ensure the quality, we filter out events that do not include the location in their name or when their *category relevance* and *query relevance* scores, provided by *EventRegistry*, are below the average scores of all events per query location. The intuition behind this step is that an event with a location mentioned in its name is more likely to provide news articles focusing on the queried location. Finally, we extract all news articles from the remaining event clusters.

Data Filtering We apply the following steps to remove irrelevant samples: *(1) Named Entity-Query Location Match:* We assume that an article is related to a query location if it is geographically close to at least one named entity. Following related work (Müller-Budack et al. 2021), we extract the named entities using *spaCy* (Honnibal et al. 2020) and use *Wikifier* (Brank et al. 2018) to link them to *Wikidata* for disambiguation. We extract *coordinate location* (*Wikidata*

[3] Source code and dataset: https://github.com/TIBHannover/mm-locate-news.

[4] http://eventregistry.org/

Property P625), which is available primarily for locations (e.g. landmarks, cities, or countries). For persons, we extract the *place of birth* (*Wikidata Property P19*) as they likely act in the respective country (or even city). We compute the Great Circle Distance (GCD) between the geographical coordinates of the query location and the extracted entity locations. We keep news articles that include at least one named entity whose GCD from the query location is smaller than $\sqrt[k]{a}$, where a is the area (*Wikidata Property P2046*) of the query location and k is a hyperparameter as defined in Sect. 2.4.2. *(2) Event-News Article Distance:* Each news article in *EventRegistry* is assigned a similarity measure that represents the closeness to an event. We discard articles with a lower similarity than the average similarity of all articles of the same cluster to keep the news articles that are most related to the respective event. *(3) Redundancy Removal:* We compute the similarity between news articles using TF-IDF vectors (Term Frequency-Inverse Document Frequency) and discard one of the articles when the similarity is higher than 0.5 to remove redundancy. *(4) Filtering of Rare Locations:* After applying filtering steps 1–3, we remove rare locations (and corresponding articles) with less than five articles as they contain too few articles for training.

Dataset Statistics In total, we queried 853 locations and extracted 13,143 news articles. After the data cleaning steps, we end up with 6,395 news articles for 389 locations (237 cities and 152 countries). We divided the *MM-Locate-News* dataset into training, validation, and test data splits by equally distributing news articles among locations as given in Table 2.3, yielding approximately 80:10:10 splits (see Fig. 2.1 for samples from the dataset).

2.3.2.2 Data Annotation Process

Data Annotation The test split of the dataset is manually annotated. Users annotated a given news article along with its image and the query location to provide “yes”, “no”, or “unsure” labels to three criteria (*C1–C3*) given in Table 2.4. Different criteria depending on the answers are turned into a different variant of the test data to evaluate the geolocation estimation models. In the *T1* version, the text focuses on the query location, and in the *T2*, both image and text represent the query location. Since it was difficult to find images where the query location is shown, we made the *T3* version where the annotators were not certain about whether the image shows

Table 2.3 Distribution of train, validation, and test samples in MM-Locate-News among continents (AF, Africa; SA, South America; EU, Europe; AS, Asia; NA, North America; OC, Oceania)

	AF	SA	EU	AS	NA	OC	Total
Train	854	216	1,604	1,842	589	161	5,266
Val	93	24	147	160	88	23	535
Test	84	29	179	202	71	26	591

Table 2.4 Manual annotation criteria (C) for the *MM-locate-News* test set variants (T). Answers "yes", "no", or "unsure" are denoted as "-", while "u" and "✓" denote "unsure" and "yes"

	T1	T2	T3
C1: Image depicts query location	-	✓	u
C2: Text focuses on query location	✓	✓	✓
C3: Image and text conceptually related	-	-	✓
Number of samples	591	65	154

the location. Thus, in cases where the text focuses on the location and the image and text are related, we assume that the image also shows the location.

Annotator Agreement A total of three users annotated the test set, where two users annotated each sample. The inter-coder agreement for the criteria *C1*, *C2*, and *C3* is 0.44, 0.38, and 0.55, respectively (according to Krippendorff's alpha Krippendorff 2011). Despite relatively moderate agreement scores, we noticed that the agreement in percent is quite high: *C2* and *C3* are 80%, and *C1* is 66.6%. This is caused by the annnotators' bias towards the answer "yes" for all criteria.

2.4 Multimodal Geolocation Estimation of Photos

We define multimodal geolocation estimation of news photos as a classification task, where the photo location is predicted based on the visual content and contextual information from the accompanied body text. The numbers of $|\mathbb{C}_g|$ locations available in the dataset for a granularity g (e.g. city, country, or continent) are considered as target classes. The $|\mathbb{C}_g|$-dimensional one-hot encoded vector $\mathbf{y}_g = \langle y_1, y_2, \ldots, y_{|\mathbb{C}_g|}\rangle \in \{0, 1\}^{|\mathbb{C}_g|}$ represents the ground-truth location. In the remainder of this section, we define the features incorporated from state-of-the-art approaches and describe the multimodal architecture and loss function.

Textual Features The pre-trained language model BERT (Bidirectional Encoder Representations from Transformers) (Devlin et al. 2019) is employed to extract two distinct types of textual features, each with 768 dimensions, from the body text of the news article. (1) We average the embeddings extracted with BERT of each sentence to create a single vector B-Bd $\in \mathbb{R}^{768}$ to encode the global contextual information. (2) To create an entity-centric embedding, denoted as B-Et $\in \mathbb{R}^{768}$, we follow related work (Müller-Budack et al. 2021) and combine *spaCy* (Honnibal et al. 2020) and *Wikifier* (Brank et al. 2018) to link location, person, and event entities to *Wikidata*. The BERT embeddings for these entities are extracted based on their *Wikidata* label. Finally, we compute the average of the entity vectors, taking into account multiple mentions of the same entity, as they may be more important for the geolocation of the photo.

Visual Features To represent the *geo-representative visual concepts*, we rely on CLIP (Contrastive Language-Image Pretraining) (Radford et al. 2021). We use ViT-B/32 image encoder to extract 512 dimensional features denoted as $\text{CLIP}_i \in \mathbb{R}^{512}$.

Network Architecture In our proposed model architecture, we aim to combine textual and visual features to predict photo geolocations on various granularities, i.e. city, country, and continent levels. Since the feature dimension of visual and textual features differ, we first encode each feature vector using l_e fully-connected (FC) layers with n_e neurons each. Next, we concatenate these embeddings and feed them into l_o output FC layers. In the hidden output layers, we use n_o neurons, and in the last output layer, the number of neurons corresponds to the number of locations $|\mathbb{C}_g|$ for a given granularity g. To leverage the hierarchical information, we employ individual classifiers for each granularity in city, country, and continent level to output probabilities $\hat{\mathbf{y}}_g \in \mathbb{R}^{|\mathbb{C}_g|}$ of size $|\mathbb{C}_{city}| = 14,331$, $|\mathbb{C}_{country}| = 241$, and $|\mathbb{C}_{continent}| = 6$. Please note that we use the *Rectified Linear Unit (ReLU)* activation function (Nair and Hinton 2010) for all layers except the last output layer that uses a *softmax*. More details are provided on GitHub.[5]

Loss Function To aggregate the granularity classifiers and highlight the hierarchical attribution, we build a multi-task learning loss function as follows:

$$\mathcal{L} = \sum_g \lambda_g \mathcal{L}_g(\mathbf{y}_g, \hat{\mathbf{y}}_g), \text{ with } g \in \{\text{city, country, continent}\}, \tag{2.1}$$

$$\mathcal{L}_g(\mathbf{y}_g, \hat{\mathbf{y}}_g) = -\mathbf{y}_g \log \hat{\mathbf{y}}_g - (1 - \mathbf{y}_g) \log(1 - \hat{\mathbf{y}}_g), \tag{2.2}$$

where λ_g are the relative weights learned during training for the different granularities, considering the difference in magnitude between losses by consolidating the log standard deviation. The cross-entropy loss $\mathcal{L}_g$ for a single granularity $g \in \{\text{city, country, continent}\}$ is defined according to (2.2).

2.4.1 Experimental Setup

This section presents the experimental setup and comparison of different architectures on both the *MMG-NewsPhoto* dataset and *BreakingNews* (Ramisa et al. 2018).

Evaluation Metrics We use the Great Circle Distance (GCD) between the geo-coordinates of the predicted and ground-truth location at several tolerable error radii (Hays and Efros 2008). These values are 25,200, and 2,500 km for city, country, and continent, respectively. Furthermore, we measure the Accuracy@k that indicates whether the ground-truth location is within the top-k model predictions.

[5] Source code and dataset: https://github.com/TIBHannover/mmg-newsphoto.

Hyperparameter Settings To extract textual features, we limit the text to 500 tokens. We set the number of FC layers to $l_e = 2$ and $l_o = 2$ and choose $n_e = 1{,}024$, $n_o = 512$ neurons. While *single-task learning* model variants (denoted with stl) are optimised using a single granularity g, the remaining models use the multi-task loss presented in (2.1) to learn from hierarchical geographical information.

Baselines We compare our models to the following baselines. Note that we did not fine-tune these models and used their official models or implementations.

$base(M, f^*)$ (Müller-Budack et al. 2018) is a state-of-the-art model for photo geolocation estimation model based on ResNet-101 (He et al. 2016) pre-trained on a subset of *YFCC100M* (Thomee et al. 2015).

$T\text{-}base(M, f^*)$ is an extension of $base(M, f^*)$, where its predictions are reduced to locations mentioned in the news body to include textual information.

$T\text{-}Freq$ is based on language models for geo-tagging text (Kordopatis-Zilos et al. 2017; Larson et al. 2017; Serdyukov et al. 2009). We employ a statistical model based on frequency of entities per city using the training set. More details are provided in the supplemental material on *GitHub*[5]. The predicted location per sample is the one with the highest probability.

VT_{CM} is based on cross-modal entity consistency of image and text (Müller-Budack et al. 2020) based on persons, locations, and events. To get predictions, we sort *Cross-modal Location Similarity (CMLS)* values and get the top k locations.

2.4.2 *Results on* MMG-NewsPhoto

Comparison of the Unimodal Models As Table 2.5 shows, regarding the visual models, CLIP_i noticeably outperforms the baseline $base(M, f^*)$ (Müller-Budack et al. 2018). Regarding the textual models, the B-Bd ⊕ B-Et surpasses the individual features. It indicates that both the contextual information and named entities and

Table 2.5 Fraction of samples (%) localised within a GCD of at most 25 km (CI, city level), 200 km (CR, country level), and 2,500 km (CT, continent level) on *MMG-NewsPhoto*

Approach	CI	CR	CT
$base(M, f^*)$ Müller-Budack et al. (2018)	10.3	20.2	40.9
CLIP_i	**30.6**	**65.5**	**78.3**
$T\text{-}Freq$	12.6	31.5	49.9
B-Bd	31.5	73.4	85.6
B-Et	31.4	73.7	83.5
B-Bd ⊕ B-Et	**32.1**	**74.7**	**84.6**
$T\text{-}base(M, f^*)$	31.2	58.8	70.7
VT_{CM} (Müller-Budack et al. 2020)	22.3	50.1	60.1
CLIP_i ⊕ B-Bd ⊕ B-Et	**43.0**	**76.7**	**83.4**

Table 2.6 Mean and median GCD divided by 1,000 km on city level for the BreakingNews test set. Models trained on *MMG-NewsPhoto* are evaluated in a zero-shot setting on *BreakingNews* and MMG → BN means that the model is finetuned on *BreakingNews*

Approach	Training	Mean	Median
$CLIP_i$	MMG	3.67	1.37
$CLIP_i$	MMG → BN	3.22	0.92
B-Bd ⊕ B-Et	MMG	2.26	**0.47**
B-Bd ⊕ B-Et	MMG → BN	**2.25**	0.51
$CLIP_i$ ⊕ B-Bd ⊕ B-Et	MMG	2.70	0.63
$CLIP_i$ ⊕ B-Bd ⊕ B-Et	MMG → BN	2.38	**0.50**
Places (Ramisa et al. 2018)	BN	3.40	**0.68**
W2V matrix (Ramisa et al. 2018)	BN	1.92	0.90
VGG19 + Places + W2V matrix (Ramisa et al. 2018)	BN	**1.91**	0.88

their frequencies play a vital role in the geolocation estimation of a news photo. Table 2.7 reports the results for Accuracy@k and shows that the $CLIP_i$ visual model is superior at the country and continent levels, but in the city level, $CLIP_i$ (stl) is slightly better. Among the textual models, the B-Bd ⊕ B-Et outperforms the rest at the country and continent levels, but it is not significantly better than B-Bd ⊕ B-Et (stl) in city level (Table 2.6).

Comparison of the Multimodal Models As presented in Table 2.5, the combination of the best unimodal features, $CLIP_i$ ⊕ B-Bd ⊕ B-Et, significantly outperforms all the other models in all granularity levels. Regarding Accuracy@k, Table 2.7 confirms the same results. For the multi-task setting, it was effective in all the granularities. In conclusion, the hierarchical information propagated from the larger granularity levels not only improves the performance at the smaller granularities, such as city, but also at the country and the continent levels.

Comparison of Different Domains Figure 2.4, right, presents the Accuracy@1 per domain for different models. As shown, the multimodal model outperforms in most of the domains. In domains like *Finance*, *Health*, and *Sports*, the visual model outperforms the textual model. In *TV show* and *World*, adding visual information does not help, and in the *Health* domain, additional textual information does not impact the performance.

Comparison of Different Concepts Figure 2.4, left, shows the Accuracy@1 per concept (see Table 2.1). As presented, the proposed multimodal model outperforms the rest in all the concepts except *public personality* and *group of people*. It is also observed that, based on the multimodal model, the concept *event* results in the lowest and *scene text* results in the highest performance.

Table 2.7 Accuracy@k (A@k) for different test sets (number of samples in brackets) of *MMG-NewsPhoto*. Approaches denoted with (stl) are trained on the respective test granularity g and do not use the multi-task loss in (2.1)

		Test_{city} (660)				$\text{Test}_{country}$(794)				$\text{Test}_{continent}$ (805)	
Approach	Modality	A@1	A@2	A@5	A@10	A@1	A@2	A@5	A@10	A@1	A@2
$base(M, f^*)$ (Müller-Budack et al. 2018)	Visual	8.3	11.2	15.9	19.8	12.7	16.9	23.2	30.1	51.1	73.7
CLIP_i (stl)	Visual	**29.1**	**38.9**	**48.5**	57.4	**61.5**	70.3	**81.0**	**88.0**	77.4	89.2
CLIP_i	Visual	27.9	37.7	**48.5**	**58.0**	**61.5**	**70.9**	80.4	85.0	**78.1**	**90.8**
$T\text{-}Freq$	Textual	10.5	14.1	19.2	24.5	31.1	38.4	48.0	54.3	55.8	70.8
B-Bd	Textual	27.9	38.2	49.2	60.2	69.5	76.2	84.3	88.7	85.0	92.8
B-Et	Textual	28.2	**40.5**	52.3	62.7	70.3	79.5	86.8	89.9	83.0	92.8
B-Bd ⊕ B-Et (stl)	Textual	**28.9**	39.2	50.9	**62.9**	70.4	78.0	86.0	91.1	83.6	**92.8**
B-Bd ⊕ B-Et	Textual	28.6	40.2	**52.9**	62.0	**70.8**	**78.6**	**87.3**	**91.2**	**84.1**	92.0
$T\text{-}base(M, f^*)$	Multimodal	27.1	36.5	43.3	44.2	62.8	74.2	79.5	80.1	75.4	86.0
VT_{CM} (Müller-Budack et al. 2020)	Multimodal	11.4	20.3	36.1	42.6	40.4	63.5	84.0	87.7	53.8	81.4
CLIP_i ⊕ B-Bd ⊕ B-Et (stl)	Multimodal	37.9	50.9	62.7	71.2	73.6	**82.2**	89.5	92.2	81.9	90.3
CLIP_i ⊕ B-Bd ⊕ B-Et	Multimodal	**39.5**	**52.1**	**64.5**	**72.7**	**73.3**	81.1	**90.1**	**92.6**	**82.9**	**92.7**

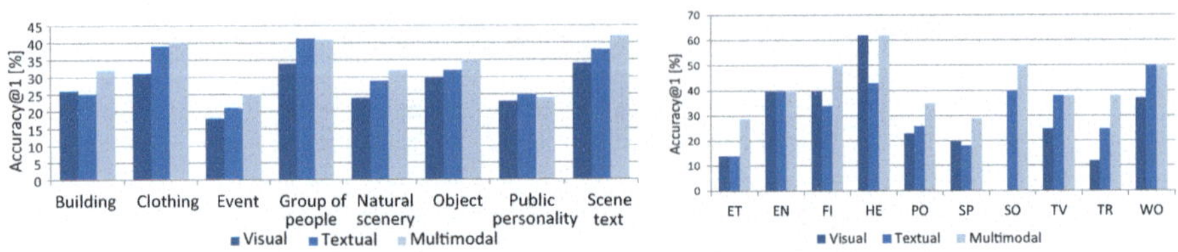

Fig. 2.4 Accuracy@1 (%) of the best-performing visual, textual, and multimodal models per concept (left) and per domain (right). ET, Entertainment; EN, Environment; FI, Finance; HE, Health; PO, Politics; SP, Sports; SO, Society; TR, Travel; TV, TV show; WO, World

Qualitative Results Figure 2.5[6] illustrates the results of different models. As expected, the visual model fails when there are only weak geo-representative concepts (Fig. 2.5a). However, it succeeds when (1) there is a strong concept (such as a landmark in Fig. 2.5b) or (2) a weak concept occurs in high frequency, e.g. *soldier* in Fig. 2.5d. The textual model fails when (1) no relevant location is mentioned (Fig. 2.5b) or (2) various irrelevant entities are mentioned, e.g. *US* in Fig. 2.5d. As expected, the textual model succeeds if there are many relevant entities to the location (Fig. 2.5a, c). When the text mentions many topics irrelevant to the image, the multimodal model fails (Fig. 2.5d). Conversely, the multimodal model succeeds in either of the following conditions: (1) the text provides rich information (both in terms of entities and content) such as Fig. 2.5a, c, or (2) the image illustrates strong visual concepts, such as Fig. 2.5b.

2.4.3 *Results on* BreakingNews

Although the image locations provided by *BreakingNews* (Ramisa et al. 2018) can be inaccurate (discussed in Sect. 2.1), we perform experiments on the dataset for comparison. *BreakingNews* includes 33,376, 11,209, and 10,580 samples for training, validation, and testing. Ramisa et al. (2018) treat the task as a regression problem where their models output the geo-coordinates. In our case, we handle the problem as a classification task to predict a specific city, country, or continent. Thus, we mapped the geo-coordinates to the closest city, country, and continent classes in *MMG-NewsPhoto* based on GCD. Table 2.6 presents the comparison of the proposed models with *BreakingNews* (abbreviated as BN) (Ramisa et al. 2018) approaches. The comparison is based on the mean and median GCD values (Ramisa et al. 2018). We evaluate our approach in two settings. In the zero-shot setting, the

[6] The photos are licenced as follows: (a) by Tryfon Topalidis and (c) by London Stock Exchange, https://creativecommons.org/licenses/by-sa/3.0/; (b) under Pixabay licence, https://pixabay.com/service/license-summary/; and (d) by NATO Training Mission-Afghanistan, https://creativecommons.org/licenses/by-sa/2.0/.

Fig. 2.5 Sample outputs from the *MMG-NewsPhoto* dataset with the predicted locations using best-performing textual, visual, and multimodal models. Predictions written in bold are correct and correspond to ground-truth (GT) locations. Images are replaced with similar ones due to licence restrictions

model was trained on *MMG-NewsPhoto* and tested on *BreakingNews* without further optimisation. In the second configuration, the best model on *MMG-NewsPhoto* is both fine-tuned and tested on *BreakingNews*. The B-Bd $\oplus$ B-Et model has the lowest median value (470 km) in the zero-shot setting and outperforms VGG19 + Places + W2V matrix (Ramisa et al. 2018) (880 km). In general, the comparison confirms the feasibility of applying the proposed models to unseen examples. In the second setting (MMG $\rightarrow$ BN), CLIP$_i$ $\oplus$ B-Bd $\oplus$ B-Et outperforms all the *BreakingNews* baselines by 180–380 km of the median value. As observed, our models perform better using the median metric, i.e. our models are better for the majority of samples.

2.5 Multimodal Focus Location Estimation of News

We define focus location estimation as a classification problem where for each article, i.e. image-text pair, the total number of n query locations (country or city) is considered as target classes. The n-dimensional one-hot encoded ground-truth vector $\mathbf{y} = \langle y_1, y_2, \ldots, y_n \rangle \in \{0, 1\}^n$ represents the query location. We extract textual and visual features as follows.

Visual Features The visual *Scene* descriptor (representing a place in a general sense) is based on ResNet-152 model (He et al. 2016) to recognise 365 places (pre-trained on the Places365 dataset Zhou et al. 2018). The *Location* descriptor, $base(M, f^*)$, is taken from a state-of-the-art photo geolocation estimation approach (Müller-Budack et al. 2018). The *Object* descriptor utilises the ResNet-152 model (He et al. 2016) pre-trained on the ImageNet dataset (Deng et al. 2009). Eventually, the $CLIP_i$ descriptor is utilised as described in Sect. 2.4.

Textual Features Regarding the textual features we use B-Bd and B-Et as described in Sect. 2.3.1.

Multimodal Architecture The textual and visual embeddings are concatenated and passed to the fully connected (FC) layers with an output size of 1,024, followed by a *ReLU* layer. Next, the outputs are fed to another *FC* layer followed by a $\tanh(u_i^l)$ layer. The norm function with clamp min $= 10^{-12}$ is applied to extract the visual $\hat{\mathbf{y}}_v$ and textual vector $\hat{\mathbf{y}}_t$ with size of $n = 389$ (number of locations). We obtain the multimodal output $\hat{\mathbf{y}}_m$ using the maximum probabilities of the visual $\hat{\mathbf{y}}_v$ and textual $\hat{\mathbf{y}}_t$ outputs. We set the parameter for data filtering to $k = 6$ (Sect. 2.5) based on an empirical evaluation of a small subset (150 samples) of the dataset.

2.5.1 Experimental Setup

In this section, we report experimental results including a comparison with state-of-the-art approaches on the *MM-Locate-News* (Sect. 2.3.2) dataset using the GCD evaluation metric (Sect. 2.4.1).

Compared Systems We evaluate different combinations of the proposed model based on the feature modalities. We also compare against two popular text-based methods, *Cliff-clavin* (D'Ignazio et al. 2014) and *Mordecai* (Halterman 2018), and one image-based state-of-the-art model (*ISNs, Individual Scene Networks* Müller-Budack et al. 2018).

2.5.2 *Results on* MM-locate-News

The results are reported in Table 2.8 and discussed below.

Textual Models For smaller GCD thresholds, specifically city and region, in *T2*, the combination *B-Et* ⊕ *B-Bd* improves the performance, and in *T1* and *T3*, the *B-Et* model provides the best results. When used separately, *B-Et* has a more substantial impact than *B-Bd*, indicating that named entities and their frequency play a vital role in predicting the focus location in the news. While *Mordecai* and *Cliff-clavin* achieve the best results at country and continent level for *T1* and *T3*, respectively, these baselines are either not applicable (*Mordecai*) or achieve worse results (*Cliff-clavin*) compared to our models on more fine-grained levels.

Visual Models The results show that $CLIP_i$ performs well in all test variants providing the best results on *T1* and *T3* and that combinations with scene ($Sc \oplus CLIP_i$) and location features ($Lo \oplus Sc \oplus CLIP_i$) can further improve the results. *ISNs* specifically trained for photo geolocalisation achieve superior results on *T2* where images depict the query location and provide enough geographical cues. Unlike $CLIP_i$, ISNs do not generalise well on other test variants.

Multimodal Models The combination of $CLIP_i$ with multimodal information drastically improves the results compared to unimodal models in all test data variants and distance thresholds. Even though our visual models do not outperform *ISNs* in *T2*, they considerably improve the results when combined with textual features ($Lo \oplus Sc \oplus B-Bd \oplus B-Et$). These results suggest that a multimodal architecture is beneficial for focus location estimation in news.

2.6 Information Retrieval

In this section, we provide a brief review of diverse methodologies that have been proposed in the realm of news retrieval. Central to our discussion is the emphasis on the significance of geographic information in news articles. Such geographic data often plays a pivotal role in tailoring and refining retrieval processes. Moreover, we investigate how multimodal information extracted from news photos and body text enhances the retrieval task.

2.6.1 *GeoWINE: Geolocation Based Wiki, Image, News, and Event Retrieval*

The proposed GeoWINE (Tahmasebzadeh et al. 2021) is a geolocation-based multimodal retrieval system that comprises five modules (see Fig. 2.6). Given an image as input, it applies a state-of-the-art geolocation estimation model as

Table 2.8 Accuracy (%) of focus location estimates for baselines and our models on the test variants of MM-Locate-News (best results per modality and GCD accuracy level are highlighted). Text features: BERT-Body (B-Bd), BERT-Entities (B-Et). Visual Features: Location (Lo), Object (Ob), Scene (Sc), $CLIP_i$. GCD accuracy levels: City (CI, max. 25 km GCD to the ground truth location), Region (RE, max. 200 km), Country (CR, max. 750 km), Continent (CT, max. 2,500 km)

Approach	Modality	T1				T2				T3			
		CI	RE	CR	CT	CI	RE	CR	CT	CI	RE	CR	CT
Mordecai (Halterman 2018) (only country-level)	Textual	–	–	**74.1**	82.9	–	–	72.3	84.6	–	–	**72.7**	81.8
Cliff-clavin (D'Ignazio et al. 2014)	Textual	36.9	53.1	71.2	**86.5**	38.5	56.9	66.2	87.7	33.1	**48.7**	**72.7**	**85.1**
B-Bd	Textual	22.8	27.4	41.1	68.4	33.8	35.4	49.2	70.8	19.5	23.4	40.9	63.6
B-Et	Textual	**48.1**	**53.5**	66.3	79.5	58.5	**64.6**	**75.4**	81.5	**42.9**	46.8	61.0	77.9
B-Bd ⊕ B-Et	Textual	42.5	47.9	60.4	78.5	**60.0**	**64.6**	73.8	**89.2**	37.0	41.6	53.2	71.4
ISNs (Müller-Budack et al. 2018)	Visual	2.5	4.4	12.0	31.0	**16.9**	**26.2**	**40.0**	**55.4**	0.6	1.9	7.8	29.9
$CLIP_i$	Visual	**4.6**	**6.3**	**15.4**	**41.1**	4.6	4.6	13.8	41.5	5.2	**8.4**	**19.5**	42.2
Lo ⊕ Ob	Visual	3.2	3.7	8.5	27.7	10.8	10.8	13.8	27.7	3.9	4.5	9.7	27.3
Sc ⊕ Ob	Visual	1.2	1.5	6.4	20.6	3.1	4.6	9.2	21.5	1.9	2.6	9.1	26.0
Lo ⊕ Sc	Visual	2.4	3.0	9.0	27.1	6.2	6.2	12.3	33.8	1.9	3.2	9.1	26.0
Lo ⊕ Sc ⊕ Ob	Visual	2.5	3.4	9.1	31.0	7.7	7.7	13.8	30.8	3.2	5.2	11.7	33.8
Lo ⊕ $CLIP_i$	Visual	3.7	5.6	12.9	36.9	4.6	7.7	13.8	41.5	5.2	7.1	16.2	37.7
Sc ⊕ $CLIP_i$	Visual	3.9	5.6	12.7	36.4	3.1	4.6	13.8	44.6	**5.8**	**8.4**	16.9	40.3
Lo ⊕ Sc ⊕ $CLIP_i$	Visual	2.5	3.7	10.5	33.8	4.6	4.6	10.8	33.8	3.2	5.2	11.7	**43.5**
$CLIP_i$ ⊕ B-Bd ⊕ B-Et	Textual+Visual	63.6	68.9	78.5	86.6	69.2	**76.9**	**84.6**	**90.8**	61.0	64.9	74.7	81.8
$CLIP_i$ ⊕ Lo ⊕ B-Bd ⊕ B-Et	Textual+Visual	61.4	66.0	76.8	86.3	66.2	70.8	81.5	**90.8**	61.0	64.9	74.0	81.8
$CLIP_i$ ⊕ Sc ⊕ B-Bd ⊕ B-Et	Textual+Visual	63.1	68.0	78.0	86.0	63.1	67.7	76.9	86.2	**63.6**	68.8	77.3	83.8
Lo ⊕ Sc ⊕ B-Bd ⊕ B-Et	Textual+Visual	65.1	69.5	78.7	84.8	70.8	75.4	81.5	86.2	**63.6**	68.2	77.9	80.5
$CLIP_i$ ⊕ Lo ⊕ Sc ⊕ B-Bd ⊕ B-Et	Textual+Visual	**65.5**	**70.6**	**81.2**	**88.7**	**72.3**	**76.9**	83.1	**90.8**	**63.6**	**69.5**	**81.2**	**85.7**

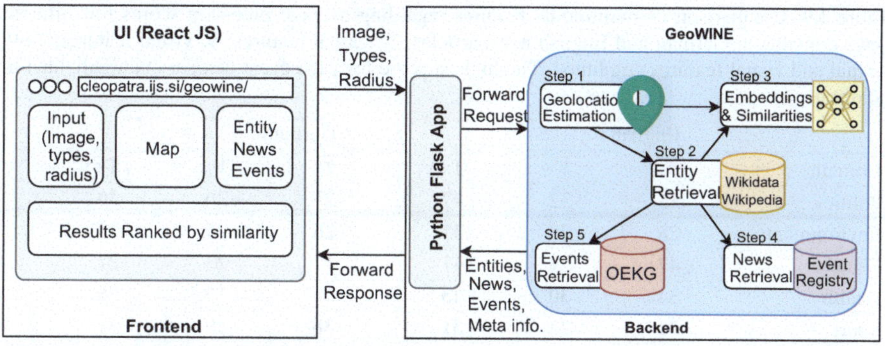

Fig. 2.6 Overview of the GeoWINE architecture

a starting point to retrieve data from Wikidata (Vrandecic and Krötzsch 2014), EventRegistry[4], and OEKG (Gottschalk et al. 2021). The geolocation estimation model predicts the coordinates of the input image. The second module performs a geospatial query on Wikidata to retrieve all entities of specific types located no farther than a specified radius from the predicted coordinates. Here, the entity types and the radius are given as input to our system. The third module leverages three different image embedding representations that are derived from the tasks of geolocation estimation and place recognition and an ImageNet model for image classification. These embeddings are used to rank the most similar entities compared to the input image. The last two modules retrieve similar news and events from EventRegistry and OEKG.

Evaluation We evaluate GeoWINE on the Google Landmarks dataset (Weyand et al. 2020), where it achieves promising performance in predicting entity labels of query images. GeoWINE enables users to retrieve entities, news, and events related to an image, through a clean and intuitive User Interface (UI), with interactive response times. To the best of our knowledge, this is the first public and open-source demo for geolocation-based multimodal retrieval through various sources. To facilitate reproducibility and reuse, all material is publicly available.[7]

Limitations Event though GeoWINE achieves promising results in geolocating landmark photos, its efficacy in predicting locations for photos in documents with a news character may be limited. To improve both geolocation and the retrieval task in the news domain, the geolocation estimation module could be replaced with multimodal approaches that combine visual features with textual information made specifically for news photos, such as the ones proposed in Sects. 2.4 and 2.5.

[7] https://github.com/cleopatra-itn/GeoWINE

Table 2.9 Comparison of multimodal features regarding average precision scores for different news domains in German and English news articles. T, textual features; V, visual features; T+V, textual and visual features combined. The highest score for each event category is highlighted in bold font

	English			German		
Domain	$\overline{T}$	$\overline{V}$	T+V	$\overline{T}$	$\overline{V}$	T+V
Politics	**55**	32	47	21	26	**30**
Environment	28	28	**37**	26	25	**33**
Finance	**47**	21	41	17	23	**27**
Health	34	30	**43**	**29**	19	28
Sport	**31**	23	**31**	**36**	19	34

2.6.2 Multimodal News Retrieval

The paper Tahmasebzadeh et al. (2020) proposes a zero-shot-based news retrieval system that uses various visual and textual features introduced in Sect. 2.5. The retrieval task is applied on a dataset including English and German news documents with 348 and 263 samples, respectively, on various domains.

Evaluation The experimental results show that regarding the combination of all features, in English news, even though visual features are not better than textual features, they helped textual features improve the overall performance for domains such as *Environment* and *Health (see T $\oplus$ V column in Table 2.9)*. On the other hand, for *Politics* and *Finance*, textual features outperform either visual and combined features. One reason is that the content of photos in these domains is not noticeable in terms of places, geolocation, or objects. The other reason is the richness of text in comparison with photos. Since these two domains include very specific events such as *Volkswagen emissions scandal* and *Greek government debt crisis*, due to specific entities existing in their texts, entity overlap outperforms the other four remaining feature types, including all visual features.

Limitations Overall, the experimental results confirm that combination of visual and textual features enhances the news retrieval task. Yet, there remains a gap in employing advanced visual descriptors to characterise the visual content of news images, particularly in domains like *Finance* and *Politics*. Making use of face detectors that can identify specific individuals depicted in photos could prove beneficial, especially given that images in these news domains frequently feature prominent figures.

2.7 Limitations and Future Work

In this chapter, we introduced various datasets and multimodal approaches for extracting geolocation of news documents (Sects. 2.5 and 2.4). In addition, we highlighted the potential of geolocalisation in information retrieval (Sects. 2.6.1 and 2.6.2).

Visual Features We have used various image descriptors to represent photos such as $CLIP_i$, *Scene*, *Location*, and *Object*. All these descriptors give one feature vector as a general representation for the whole photo. To have a better multimodal representation that is able to match various aspects of a photo to text, individual concepts in news photos, such as person, event type, and object could be represented. Furthermore, structured features could be extracted from the photo, for instance, a scene graph that represents the relation of event arguments.

Textual Features Regarding the textual features, we rely on *spaCy* (Honnibal et al. 2020), *Wikifier* (Brank et al. 2018), and BERT (Devlin et al. 2019) embeddings to extract two types of features *B-Bd* and *B-Et* as a single vectors per type. To enhance the representation of the body text of news, external knowledge such as knowledge graph information (e.g. entity type, event arguments, event dates) could be included. In addition, contextual information from image and/or text could be very impactful in news retrieval, such as event arguments and roles, news topic, and sentiment.

Multilinguality Currently, the proposed multimodal geolocation estimation models are limited to English language. In order to generalise to more languages, it is required to include named-entity recognition tools together with text encoders in the respective languages. Alternatively, a text translation tool could be integrated into the system to convert the input text in any language to English.

Applications We introduced information retrieval systems that benefit from geolocation estimation of photos (Tahmasebzadeh et al. 2021, 2020). As a future direction, the impact of geolocalisation of news documents could be investigated in various tasks in the news domain, such as fake news detection or news recommendation based on location desires. On the other hand, the proposed systems for geolocalisation of photos could be integrated in the OEKG (Gottschalk et al. 2021), for instance, to extend the nodes with the corresponding images or to connect visually similar entities to the nodes based on geolocation.

Acknowledgments The work presented in this chapter has received funding from the European Union's Horizon 2020 research and innovation program under the Marie Skłodowska-Curie grant agreement no. 812997 and under the name CLEOPATRA (Cross-lingual Event-centric Open Analytics Research Academy)

References

Armitage J, Kacupaj E, Tahmasebzadeh G, Swati, Maleshkova M, Ewerth R, Lehmann J (2020) MLM: a benchmark dataset for multitask learning with multiple languages and modalities. In: International Conference on Information and Knowledge Management, CIKM, ACM, pp 2967–2974. https://doi.org/10.1145/3340531.3412783

Avrithis Y, Kalantidis Y, Tolias G, Spyrou E (2010) Retrieving landmark and non-landmark images from community photo collections. In: International Conference on Multimedia, MM, ACM, pp 153–162. https://doi.org/10.1145/1873951.1873973

Baatz G, Saurer O, Köser K, Pollefeys M (2012) Large scale visual geo-localization of images in mountainous terrain. In: European Conference on Computer Vision, ECCV, Springer, pp 517–530. https://doi.org/10.1007/978-3-642-33709-3_37

Berton GM, Masone C, Caputo B (2022) Rethinking visual geo-localization for large-scale applications. In: Conference on Computer Vision and Pattern Recognition, CVPR, IEEE, pp 4868–4878. https://doi.org/10.1109/CVPR52688.2022.00483

Biten AF, Gómez L, Rusiñol M, Karatzas D (2019) Good news, everyone! context driven entity-aware captioning for news images. In: Conference on Computer Vision and Pattern Recognition, CVPR, Computer Vision Foundation/IEEE, pp 12466–12475. https://doi.org/10.1109/CVPR.2019.01275

Boiarov A, Tyantov E (2019) Large scale landmark recognition via deep metric learning. In: Zhu W, Tao D, Cheng X, Cui P, Rundensteiner EA, Carmel D, He Q, Yu JX (eds) ACM International Conference on Information and Knowledge Management, CIKM, ACM, pp 169–178. https://doi.org/10.1145/3357384.3357956

Brank J, Leban G, Grobelnik M (2018) Semantic annotation of documents based on wikipedia concepts. Informatica (Slovenia) 42(1):23–32

Brejcha J, Cadík M (2017) State-of-the-art in visual geo-localization. Pattern Analy Appl 20(3):613–637. https://doi.org/10.1007/s10044-017-0611-1

Cheng J, Wu Y, AbdAlmageed W, Natarajan P (2019) QATM: quality-aware template matching for deep learning. In: Conference on Computer Vision and Pattern Recognition, CVPR, Computer Vision Foundation/IEEE, pp 11553–11562. https://doi.org/10.1109/CVPR.2019.01182

Crandall DJ, Backstrom L, Huttenlocher DP, Kleinberg JM (2009) Mapping the world's photos. In: International Conference on World Wide Web, WWW, ACM, pp 761–770. https://doi.org/10.1145/1526709.1526812

Deng J, Dong W, Socher R, Li LJ, Li K, Fei-Fei L (2009) ImageNet: a large-scale hierarchical image database. In: IEEE Conference on Computer Vision and Pattern Recognition, pp 248–255

Devlin J, Chang M, Lee K, Toutanova K (2019) BERT: pre-training of deep bidirectional transformers for language understanding. In: Conference of the North American Chapter of the Association for Computational Linguistics: Human Language Technologies, NAACL-HLT, Association for Computational Linguistics, pp 4171–4186. https://doi.org/10.18653/v1/n19-1423

D'Ignazio C, Bhargava R, Zuckerman E, Beck L (2014) Cliff-clavin: determining geographic focus for news articles. In: NewsKDD:Data Science for News Publishing Workshop Co-located with ACM SIGKDD Conference on Knowledge Discovery and Data Mining

Gottschalk S, Kacupaj E, Abdollahi S, Alves D, Amaral G, Koutsiana E, Kuculo T, Major D, Mello C, Cheema GS, Sittar A, Swati, Tahmasebzadeh G, Thakkar G (2021) OEKG: the open event knowledge graph. In: International Workshop on Cross-lingual Event-centric Open Analytics co-located with the 30th The Web Conference (WWW), CEUR-WS.org, CEUR Workshop Proceedings, vol 2829, pp 61–75

Halterman A (2018) Linking Events and Locations in Political Text. https://doi.org/10.2139/ssrn.3267476. MIT Political Science Department Research Paper

Hays J, Efros AA (2008) IM2GPS: estimating geographic information from a single image. In: Conference on Computer Vision and Pattern Recognition, CVPR, IEEE Computer Society. https://doi.org/10.1109/CVPR.2008.4587784

He K, Zhang X, Ren S, Sun J (2016) Deep residual learning for image recognition. In: Conference on Computer Vision and Pattern Recognition, CVPR, IEEE Computer Society, pp 770–778. https://doi.org/10.1109/CVPR.2016.90

Honnibal M, Montani I, Van Landeghem S, Boyd A (2020) spaCy: Industrial-strength natural language processing in Python. https://zenodo.org/doi/10.5281/zenodo.1212303

Izbicki M, Papalexakis EE, Tsotras VJ (2019) Exploiting the earth's spherical geometry to geolocate images. In: European Conference on Machine Learning and Knowledge Discovery in Databases, ECML PKDD, Springer, pp 3–19. https://doi.org/10.1007/978-3-030-46147-8_1

Kim HJ, Dunn E, Frahm J (2017) Learned contextual feature reweighting for image geo-localization. In: Conference on Computer Vision and Pattern Recognition, CVPR, IEEE Computer Society, pp 3251–3260. https://doi.org/10.1109/CVPR.2017.346

Kordopatis-Zilos G, Popescu A, Papadopoulos S, Kompatsiaris Y (2016) Placing images with refined language models and similarity search with PCA-reduced VGG Features. In: MediaEval 2016 Workshop, CEUR-WS.org

Kordopatis-Zilos G, Papadopoulos S, Kompatsiaris I (2017) Geotagging text content with language models and feature mining. Proc IEEE 105(10):1971–1986. https://doi.org/10.1109/JPROC.2017.2688799

Kordopatis-Zilos G, Galopoulos P, Papadopoulos S, Kompatsiaris I (2021) Leveraging efficientnet and contrastive learning for accurate global-scale location estimation. In: International Conference on Multimedia Retrieval, ICMR, ACM, pp 155–163. https://doi.org/10.1145/3460426.3463644

Krippendorff K (2011) Computing Krippendorff's alpha-reliability. https://repository.upenn.edu/asc_papers/43

Larson MA, Soleymani M, Gravier G, Ionescu B, Jones GJF (2017) The benchmarking initiative for multimedia evaluation: mediaeval 2016. IEEE MultiMedia 24(1):93–96

Mackenzie JM, Benham R, Petri M, Trippas JR, Culpepper JS, Moffat A (2020) CC-News-En: a large english news corpus. In: International Conference on Information and Knowledge Management, CIKM, ACM, pp 3077–3084. https://doi.org/10.1145/3340531.3412762

Müller-Budack E, Pustu-Iren K, Ewerth R (2018) Geolocation estimation of photos using a hierarchical model and scene classification. In: European Conference on Computer Vision, ECCV, Springer, pp 575–592. https://doi.org/10.1007/978-3-030-01258-8_35

Müller-Budack E, Theiner J, Diering S, Idahl M, Ewerth R (2020) Multimodal analytics for real-world news using measures of cross-modal entity consistency. In: International Conference on Multimedia Retrieval, ICMR, ACM, pp 16–25. https://doi.org/10.1145/3372278.3390670

Müller-Budack E, Theiner J, Diering S, Idahl M, Hakimov S, Ewerth R (2021) Multimodal news analytics using measures of cross-modal entity and context consistency. Int J Multimedia Inf Retr 10(2):111–125. https://doi.org/10.1007/s13735-021-00207-4

Nair V, Hinton GE (2010) Rectified linear units improve restricted Boltzmann machines. In: Fürnkranz J, Joachims T (eds) International Conference on Machine Learning (ICML), Omnipress, pp 807–814

Radford A, Kim JW, Hallacy C, Ramesh A, Goh G, Agarwal S, Sastry G, Askell A, Mishkin P, Clark J, Krueger G, Sutskever I (2021) Learning transferable visual models from natural language supervision. In: International Conference on Machine Learning, ICML, PMLR, pp 8748–8763

Ramisa A, Yan F, Moreno-Noguer F, Mikolajczyk K (2018) BreakingNews: article annotation by image and text processing. IEEE Trans Pattern Analy Mach Intell 40(5):1072–1085. https://doi.org/10.1109/TPAMI.2017.2721945

Seo PH, Weyand T, Sim J, Han B (2018) CPlaNet: enhancing image geolocalization by combinatorial partitioning of maps. In: European Conference on Computer Vision, ECCV, Springer, pp 544–560. https://doi.org/10.1007/978-3-030-01249-6_33

Serdyukov P, Murdock V, van Zwol R (2009) Placing flickr photos on a map. In: SIGIR Conference on Research and Development in Information Retrieval, SIGIR, ACM, pp 484–491. https://doi.org/10.1145/1571941.1572025

Singhal S, Shah RR, Chakraborty T, Kumaraguru P, Satoh S (2019) SpotFake: a multi-modal framework for fake news detection. In: IEEE International Conference on Multimedia Big Data, BigMM, IEEE, pp 39–47. https://doi.org/10.1109/BigMM.2019.00-44

Tahmasebzadeh G, Hakimov S, Müller-Budack E, Ewerth R (2020) A feature analysis for multimodal news retrieval. In: Demidova E, Hakimov S, Winters J, Tadic M (eds) 1st International Workshop on Cross-lingual Event-centric Open Analytics co-located with the 17th Extended Semantic Web Conference (ESWC 2020), Heraklion, Crete, June 3, 2020 (Online event due to COVID-19 outbreak), CEUR-WS.org, CEUR Workshop Proceedings, vol 2611, pp 43–56

Tahmasebzadeh G, Kacupaj E, Müller-Budack E, Hakimov S, Lehmann J, Ewerth R (2021) GeoWINE: geolocation based wiki, image, news and event retrieval. In: Diaz F, Shah C, Suel T, Castells P, Jones R, Sakai T (eds) SIGIR '21: The 44th International ACM SIGIR Conference on Research and Development in Information Retrieval, Virtual Event, July 11–15, 2021, ACM, pp 2565–2569. https://doi.org/10.1145/3404835.3462786

Tahmasebzadeh G, Müller-Budack E, Hakimov S, Ewerth R (2022) MM-Locate-News: Multimodal Focus Location Estimation in News. arXiv preprint arXiv:221108042

Tahmasebzadeh G, Hakimov S, Ewerth R, Müller-Budack E (2023) Multimodal geolocation estimation of news photos. In: European Conference on Information Retrieval, ECIR, Springer, pp 204–220

Theiner J, Müller-Budack E, Ewerth R (2022) Interpretable semantic photo geolocation. In: Winter Conference on Applications of Computer Vision, WACV, IEEE, pp 1474–1484. https://doi.org/10.1109/WACV51458.2022.00154

Thomee B, Shamma DA, Friedland G, Elizalde B, Ni K, Poland D, Borth D, Li L (2015) The new data and new challenges in multimedia research. CoRR abs/1503.01817. 1503.01817

Tomesek J, Cadík M, Brejcha J (2022) CrossLocate: cross-modal large-scale visual geolocalization in natural environments using rendered modalities. In: Winter Conference on Applications of Computer Vision, WACV, IEEE, pp 2193–2202. https://doi.org/10.1109/WACV51458.2022.00225

Trevisiol M, Jégou H, Delhumeau J, Gravier G (2013) Retrieving geo-location of videos with a divide & conquer hierarchical multimodal approach. In: International Conference on Multimedia Retrieval, ICMR, ACM, pp 1–8. https://doi.org/10.1145/2461466.2461468

Uzkent B, Sheehan E, Meng C, Tang Z, Burke M, Lobell DB, Ermon S (2019) Learning to interpret satellite images using wikipedia. In: International Joint Conference on Artificial Intelligence, IJCAI, ijcai.org, pp 3620–3626

Vo NN, Jacobs N, Hays J (2017) Revisiting IM2GPS in the deep learning era. In: International Conference on Computer Vision, ICCV, IEEE Computer Society, pp 2640–2649

Vrandecic D, Krötzsch M (2014) Wikidata: a free collaborative knowledgebase. Commun ACM 57(10):78–85. https://doi.org/10.1145/2629489

Weyand T, Kostrikov I, Philbin J (2016) PlaNet - photo geolocation with convolutional neural networks. In: European Conference on Computer Vision, ECCV, Springer, pp 37–55. https://doi.org/10.1007/978-3-319-46484-8_3

Weyand T, Araujo A, Cao B, Sim J (2020) Google landmarks dataset v2 - a large-scale benchmark for instance-level recognition and retrieval. In: Conference on Computer Vision and Pattern Recognition, CVPR, IEEE, pp 2572–2581. https://doi.org/10.1109/CVPR42600.2020.00265

Zhou B, Lapedriza À, Khosla A, Oliva A, Torralba A (2018) Places: a 10 million image database for scene recognition. IEEE Trans Pattern Analy Mach Intell 40(6):1452–1464. https://doi.org/10.1109/TPAMI.2017.2723009

Chapter 3
Robustness of Corpus-Based Typological Strategies for Dependency Parsing

Diego Alves and Daniel Gomes

Abstract This chapter presents a comparison of the corpus-based typological classification of ten European Union languages obtained using parallel corpora, with one generated in a less controlled scenario, with non-parallel automatically annotated data. First, we described the specific pipeline that was created to extract and annotate multilingual data from the Arquivo.pt 2019 European Parliamentary Elections collection. Two new corpora for all EU languages were generated and made publicly and freely available: one composed of raw texts extracted from this collection and the other with syntactic annotation obtained automatically. Then, we presented an overview of different quantitative typological approaches developed for dependency parsing improvement and selected the most optimised ones to conduct our comparative analysis. Finally, we compared both scenarios using the same corpus-based strategy and showed that the classification obtained using the data provided by the Arquivo.pt dataset provides valuable linguistic information for this type of study, presenting similarities when compared to the classification based on parallel corpora. However, considering the dissimilarities observed, further analysis is required before validating this new method.

D. Alves (✉)
Faculty of Humanities and Social Sciences, University of Zagreb, Zagreb, Croatia
e-mail: dfvalio@ffzg.hr

D. Gomes
FCT-FCCN, Lisboa, Portugal
e-mail: daniel.gomes@fccn.pt

I. Marenzi et al. (eds.), *Event Analytics across Languages and Communities*,
https://doi.org/10.1007/978-3-031-64451-1_3

3.1 Introduction

Dependency parsing is a natural language processing (NLP) task that determines the grammatical structure of a sentence by examining the syntactic relations between its linguistic units. In other words, it consists of identification of heads and dependents as well as the type of relationship between them (Jurafsky and Martin 2009)

Since the 1980s, dependency parsing tools have increasingly relied on statistics, probability and machine learning methods, which require a large amount of linguistic data. Furthermore, from 2015 onward, the usage of deep learning techniques has been dominant in this NLP field. These approaches require a large amount of annotated data, which can be problematic for some languages considered low resourced (Otter et al. 2018).

Linguistic manual annotation of texts can be very costly, and therefore, other solutions for improving dependency parsing tagging scores have been proposed over the years. One way to overcome this issue is to combine data from similar languages according to established typological classifications. Many studies have been carried out regarding typological strategies based on well-known databases that provide valuable typological information for language comparison (Ponti et al. 2019). Another efficient possibility is to conduct corpus-based typological analyses of different languages by extracting specific syntactic patterns from annotated corpora. It has been shown that these quantitative methods can be used to classify languages in an optimised way for dependency parsing improvement when parallel corpora are compared (Alves et al. 2023).

While the usage of typological databases can be problematic for languages that are not well characterised in the literature, strategies based on the extraction of syntactic information from annotated parallel data can present some limitations regarding the availability of parallel tree-banks. Moreover, some languages may have insufficient manually annotated data for conducting the quantitative analysis of the syntactic patterns. Thus, it is crucial to check how the corpus-based approaches work in a less controlled scenario (i.e. with non-parallel corpora annotated automatically using a dependency parsing tool).

The idea of the study presented in this chapter is to compare the corpus-based syntactic classifications of languages obtained via the analysis of the manually annotated parallel corpora (i.e. parallel universal dependency corpora, PUD, Zeman et al. 2018) presented by Alves et al. (2023) with those generated via the extraction of syntactic patterns from non-parallel corpora automatically annotated.

For this aim, the aforementioned classifications of the 10 European Union languages present in the PUD collection are compared to the ones obtained from the analysis of automatically annotated texts from the 2019 European Parliamentary Elections collection from the Arquivo.pt[1] (i.e. the Portuguese Web archive, a research infrastructure that preserves historical Web content Gomes 2022). The ID

[1] https://arquivo.pt/ee2019/

of this collection is EAWP23, and it was generated according to the terminology adopted by Arquivo.pt.[2] These annotated texts were obtained with a specific pipeline for optimally extracting and annotating raw text.

The chapter is composed as follows. In Sect. 3.2, we present related work regarding typological approaches for dependency parsing improvement. Then, Sect. 3.3 presents the 2019 European Parliamentary Elections collection of the Arquivo.pt together with the pipeline developed for extraction and annotation of the texts. In Sect. 3.4, we detail how different typological methods can be used for dependency parsing improvement and present the optimised corpus-based syntactic classification obtained with the PUD corpora. Afterward, in Sect. 3.5, we compare the typological results obtained with the automatically annotated corpora from the Arquivo.pt collection with the one established in the literature. Finally, Sect. 3.6 presents the conclusions and perspectives for future research.

3.2 Related Work

The WALS database is one of the most used typological resources in NLP studies (Ponti et al. 2019). It contains phylogenetic, phonological, morphosyntactic and lexical information for a large number of languages that can be used for a large variety of linguistic studies (Dryer and Haspelmath 2013). Along with that, the URIEL Typological Compendium was conceived as a meta-repository that is composed of numerous databases (WALS included) and is the base of the lang2vec tool (Littell et al. 2017). This tool is a powerful resource that allows languages to be characterised as vectors composed of typological features associated with specific values. Users can choose the type of features (e.g. genealogical, phonological, syntactic, etc.) according to their precise needs. While proposing an effective way to compare languages typologically, this tool does not characterise all languages homogeneously as it depends on the availability of linguistic descriptions provided by its sources. Thus, low-resourced languages usually have less information. For example, it is not possible to compare all 24 European Union official languages as there are no common features with valid values for all of them. Furthermore, typological databases usually fail to illustrate the variations that can occur within a single language (i.e. in general, only the most frequent phenomena are reported in the literature, not all attested ones).

In terms of corpus-based typological studies, a broad survey was provided by Levshina (2022). The author showed that while several authors quantitatively analysed specific word-order patterns (e.g. subject, verb and object position Östling 2015 and verb and locative phrases Wälchli 2009), other researchers have focused on quantitative analyses regarding language complexity (e.g. Hawkins 2003 and Sinnemäki 2014). On the other hand, the concept of typometrics was introduced

[2] https://arquivo.pt/collections

by Gerdes et al. (2021). The focus of their research was to extract rich details from corpora for testing typological implicational universals and explore new kinds of universals, named quantitative ones. Thus, different word-order phenomena were analysed quantitatively (i.e. the distribution of their occurrences in annotated corpora) to identify the ones present in all or most languages.

Thus, it is possible to notice that most studies regarding quantitative typology focus either on the analysis of specific linguistic phenomena or on the identification of universals. Our approach differs from theirs as our aim is to compare languages (i.e. language vectors) using quantitative information concerning all syntactic structures extracted from corpora to obtain a more general syntactic overview of the elements in our language set and use the results as strategies to improve dependency parsing results.

An interesting method concerning the extraction and comparison of syntactic information from tree-banks was developed by Blache et al. (2016). The MarsaGram tool is a resource that allows syntactic information (together with its statistics) to be extracted from annotated corpora by inferring context-free grammars from the syntactic structures. MarsaGram allows the extraction of linear patterns (i.e. if a specific part-of-speech precedes another one inside the same sub-tree ruled by a determined head). The authors conducted a cluster analysis comparing ten different languages and showed the potential in terms of typological analysis of this resource. However, the results were only compared to the genealogical classification of the selected languages and did not provide any comparison to other corpus-based methods. Moreover, the authors did not use the obtained classification from the perspective of improving dependency parsing systems via corpora combination.

One example of effective usage of typological features (from the URIEL database) to improve results of NLP methods was presented by Üstün et al. (2020). The authors developed the UDapter tool that uses a mix of automatically curated and predicted typological features as direct input to a neural parser. The results showed that this method allows the improvement of the dependency parsing accuracy for low-resourced languages. A similar study, using a different deep-learning architecture, was conducted by Ammar et al. (2016); however, in both cases, there is no detailed analysis of which features were the most relevant.

Furthermore, Lynn et al. (2014) proposed a study concerning the Irish language using delexicalised corpora. The authors performed a series of cross-lingual direct transfer parsing for the Irish language, and the best results were achieved with a model trained with the Indonesian corpus, a language from the Austronesian language family. The authors proposed some analysis considering similarities between the tree-banks of both languages in terms of dependency parsing labels, but a detailed statistical analysis of corpora and a complete comparison of specific typological features were not carried out.

While some papers focus on genealogical features, others consider syntactic ones. For example, Alzetta et al. (2020) presented a study whose aim was to identify cross-lingual quantitative trends in the distribution of dependency relations in annotated corpora from distinct languages by using the algorithm LISCA—LInguiStically-driven Selection of Correct Arcs (Dell'Orletta et al. 2013)—which detects patterns of syntactic structures in tree-banks. Only four Indo-European languages were scrutinised, but some interesting insights concerning language peculiarities were observed.

3.3 The 2019 European Parliamentary Elections Collection of the Arquivo.pt

As previously mentioned, in this section, the 2019 European Parliamentary Elections collection of the Arquivo.pt is presented in detail together with the process of extracting and annotating text that was developed to generate the corpora that are used in the corpus-based typological analyses.

3.3.1 Collection Description

The 2019 European Parliamentary Elections are an event of international relevance. The strategy adopted to preserve the World Wide Web has been to delegate national institutions the responsibility of selecting and preserving information relevant to their hosting countries. However, the preservation of Web pages that document transnational events is not officially assigned. Thus, the Arquivo.pt team, with the aim of preserving this specific Web content, applied a combination of human and automatic selection processes, maximising the coverage of this cross-lingual event, to guarantee the conservation of this content.

The process of generating the 2019 European Parliamentary Elections involved two main steps:

- Semi-automatic selection of relevant online content
- Crawling Web content

In the first step, 40 relevant terms in Portuguese about the 2019 European Parliamentary Elections were identified and then automatically translated into the 24 official languages of the European Union.[3] These translations were reviewed in collaboration with the Publications Office of the European Union.[4] The list of the terms in English is presented in Table 3.1.

Then, these terms were used to automatically query a Web search engine to get a total of 12,147 URLs to seed the crawls. This automation of the selection process

---

[3] Bulgarian, Croatian, Czech, Danish, Dutch, English, Estonian, Finnish, French, German, Greek, Hungarian, Irish, Italian, Latvian, Lithuanian, Maltese, Polish, Portuguese, Romanian, Slovak, Slovenian, Spanish and Swedish

[4] https://op.europa.eu/en/home

Table 3.1 List of relevant terms identified for the creation of the European Parliamentary Elections collection. The terms in English were translated into all the EU official languages in collaboration with the Publications Office of the European Union

Relevant terms
Elections European Parliament 2019
European elections 2019
European abstention 2019
Electoral Results 2019
European winners 2019
Brexit European Elections 2019
2019 European Elections
European Elections Blogs
social media European elections
European Twitter elections
European Facebook Elections
Comics European Elections 2019
European Survey 2019
European television 2019
European television debate
European caravan 2019
European representative 2019
European policy 2019
European Elections 2019 Accessibility
European Elections 2019 Parity
European candidates 2019
European deputies 2019
European 2019 Environmentalists
European elections 2019 Immigration
European List 2019
European Defeated 2019
European right 2019
European Liberals 2019
far-right European
European left 2019
European coalition
European Electoral System 2019
future of Europe 2019
European FakeNews 2019
Europeist Elections 2019
anti-Europeist 2019
European Eurochists 2019
European ideology 2019
2019 European campaign financing
MEP

enabled the expansion of information coverage about the event to multiple countries and languages without significantly increasing the number of resources required.

In parallel, a collaborative list was launched to gather contributions of relevant seeds from the international community. This initiative was disseminated through Portuguese and international contacts, like Arquivo.pt social media or the International Internet Preservation Consortium mailing lists. A total of 608 contributions were received from 16 countries. Slovakia and Portugal were the countries that suggested the highest number of seeds (114). These contributions were classified into the following categories: generic links, official sites, blogs and opinion articles, independent and/or commercial channels, social media news, satire, social network links, political parties and associations, candidates, EU institutions, agencies and bodies and European Elections 2019 News.

Regarding the second step, the Arquivo.pt team iteratively ran six crawls using different configurations and crawling software (Heritrix 3.3.0,[5] Brozzler[6] and Browsertrix[7]) to maximise the quality of the collected content. One crawl was executed before the elections and five afterward. These crawls were performed between May and July 2019 and resulted in the collection of 99 million URLs (4.8 TB).

This Web data was aggregated into one special collection identified as EAWP23 and became searchable and accessible through Arquivo.pt in July 2020.[8] Moreover, this collection is also available for automatic processing through the Arquivo.pt API[9], which was used for the text extraction presented in this study.

3.3.2 *Text Extraction and Annotation*

The schema presented in Fig. 3.1 describes the pipeline developed to extract and annotate the texts from the 2019EUElections collection. Each step is detailed in this subsection. This process was applied to each European Union official language, generating two collections of corpora: raw and annotated texts following the CoNLL-U format described by the Universal Dependencies framework.[10]

The first step of the process concerns the collection of the list of URLs (json file) concerning the list of terms in relation to the 2019 European Parliamentary Elections. For each language, we used the automatically translated version obtained

[5] https://github.com/internetarchive/heritrix3

[6] https://github.com/internetarchive/brozzler

[7] https://github.com/webrecorder/browsertrix-crawler

[8] https://arquivo.pt/ee2019

[9] https://github.com/arquivo/pwa-technologies/wiki/Arquivo.pt-API

[10] https://universaldependencies.org/

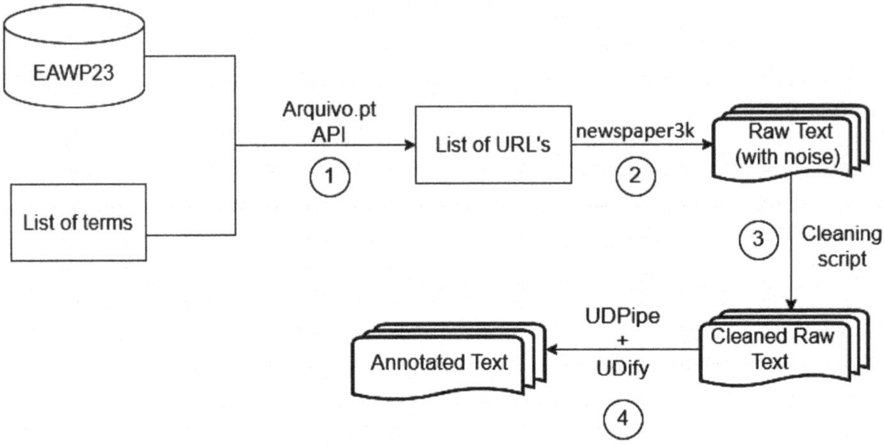

Fig. 3.1 Text extraction and annotation process

from the Portuguese original list, which was created by Arquivo.pt for the creation of the 2019EUElections collection.[11]

Then, in the second step, we extracted, for each language, the text from each URL (i.e. "linkToOriginalFile" entries from the json files) using the newspaper3k Python library.[12] Moreover, for each extracted text from the URLs, we verified the language using the langdetect 1.0.9 Python library.[13] The aim of this verification was to avoid texts written in different languages inside each specific corpus.[14]

A qualitative analysis of the corpora obtained in the second step showed that a cleaning step was necessary as the extracted texts presented some noise (e.g. instances formed by numbers, URLs, lists of names, etc.). Thus, in the third step, we applied a Python script that excluded:

- repeated instances
- instances starting with "www" or "@"
- instances composed only of capital letters
- empty lines
- instances with a ratio between characters and numbers lower than 0.95

[11] An example of a query with an English term is: https://arquivo.pt/textsearch?q="European winners2019"&prettyPrint=true&maxItems=2000.

[12] https://newspaper.readthedocs.io/en/latest/

[13] https://pypi.org/project/langdetect/

[14] The langdetect library cannot identify Maltese or Irish; thus, for these languages, we verified if the text was not written in English in the second step.

In the third step, we again verified the language for each instance as, in the previous step, the whole text of each URL was checked; thus, some specific sentences in other languages could still be present in each corpus.[15]

Therefore, we generated a collection of corpora composed of texts regarding the 2019 European Parliamentary Elections for all European Union official languages.[16] As it is composed of texts regarding a specific political event from 24 different languages and a large variety of news and other media sources, it is a useful resource for cross-lingual studies in Digital Humanities and many NLP fields.

Figure 3.2 presents a description of each corpus composing this collection, giving the number of sentences and tokens.

As expected, the English corpus is the largest (i.e. more than 32 million tokens). Other languages usually described as well resourced also have a considerably large corpus. This is the case for French, Portuguese, Spanish and German. Maltese and Irish are the two languages with the smallest corpus (301,196 and 558,031 tokens, respectively). Texts were concatenated in a randomised way, and no metadata is provided.

With the corpora described above composed of raw texts, we proceeded with the automatic annotation regarding part-of-speech and dependency relations (step 4). For this aim, we used the UDify tool (Kondratyuk and Straka 2019), which proposes an architecture aimed at part-of-speech and dependency parsing tagging integrating the Multilingual BERT language model (104 languages) (Pires et al. 2019). This tool was selected as it presents state-of-the-art algorithms concerning the specific task of dependency parsing annotation. As UDify does not propose a tokenisation module, as a pre-processing step, we tokenised the texts using UDPipe 1.0 (Straka et al. 2016).

For the automatic annotation of the texts, we applied the multilingual model proposed by Kondratyuk and Straka (2019). This model is presented as an optimised one with an increased labelled attachment score (LAS) for low-resourced languages. LAS is the standard measure for dependency parsing tools and corresponds to the percentage of words that are assigned both the correct syntactic head and the correct dependency label.[17]

The texts in the annotated corpora follow the same randomised order as the respective raw text files. Each sentence is annotated following the CoNNL-U format; thus, for each token, the following information is provided:[18]

- ID: Word index, integer starting at 1 for each new sentence
- FORM: Word form or punctuation symbol
- LEMMA: Lemma or stem of word form
- UPOS: Universal part-of-speech tag

---

[15] For Maltese and Irish, we verified if the specific character of these languages was present in the instance ([ħżġċ] for Maltese and [áéíóúÁÉÍÓÚ] for Irish).

[16] https://dataverse.harvard.edu/dataset.xhtml?persistentId=doi:10.7910/DVN/WCGNHU

[17] http://universaldependencies.org/conll17/evaluation.html

[18] https://universaldependencies.org/format.html

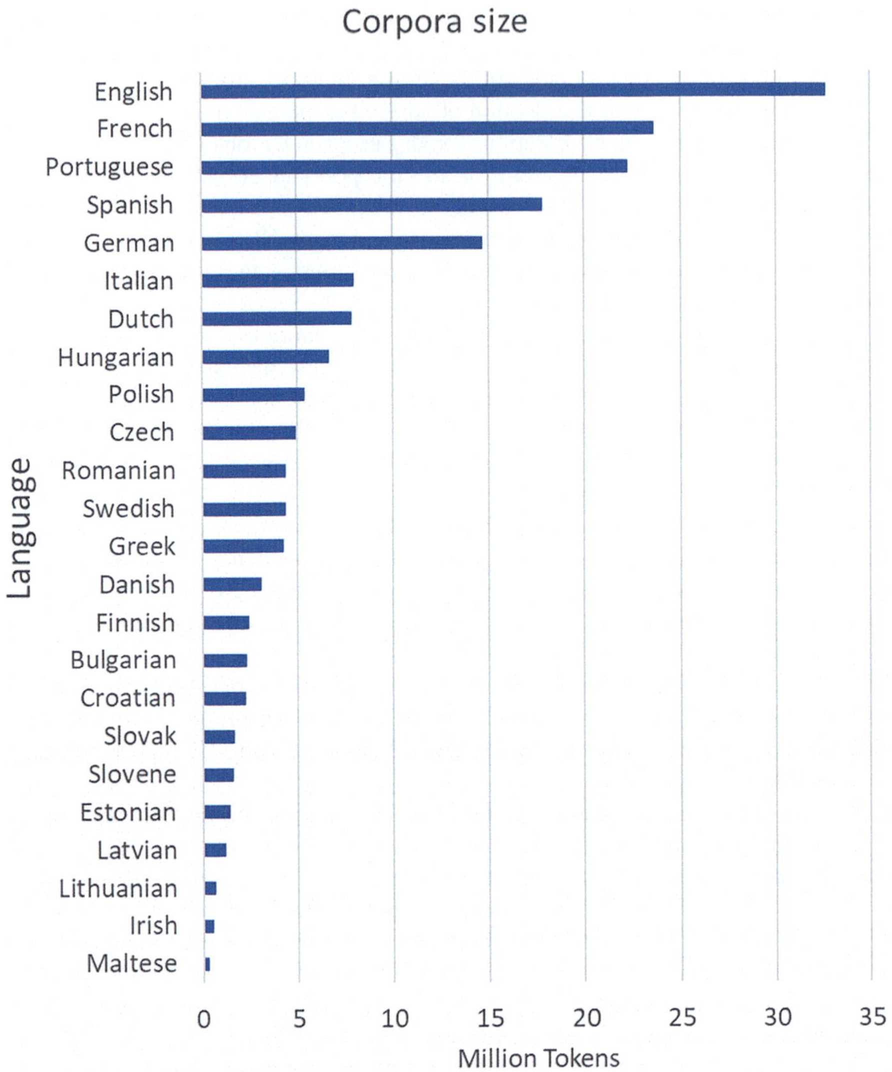

Fig. 3.2 Number of tokens of each corpus extracted from the collection 2019EUElections of the Arquivo.pt

Table 3.2 LAS scores for each language in the 2019EUElections as presented by Kondratyuk and Straka (2019) using UDify (multilingual model) and Universal Dependencies corpora as test-sets

Language	LAS
Polish	94.6
Slovak	93.9
Italian	93.7
Slovenian	93.1
Czech	92.9
Portuguese	92.5
Bulgarian	92.4
Greek	92.2
French	91.5
Dutch	91.2
Spanish	90.5
English	90.1
Croatian	89.8
Swedish	89.0
Romanian	88.6
Estonian	86.7
Finnish	86.6
Latvian	85.1
Hungarian	84.9
Danish	84.5
German	84.5
Maltese	75.6
Lithuanian	69.3
Irish	69.3

- XPOS: Language-specific part-of-speech tag; underscore if not available
- FEATS: List of morphological features from the universal feature inventory
- HEAD: Head of the current word, which is either a value of ID or zero (0)
- DEPS: Enhanced dependency graph in the form of a list of head-deprel pairs[19]
- MISC: Any other annotation

As previously explained, the idea is to test the different quantitative typological approaches using the automatically annotated texts as the base to extract the different syntactic patterns. The aim is to test the robustness of the corpus-based approaches using corpora with dependency parsing annotation containing some errors due to the automatic nature of the process. Table 3.2 presents the best LAS results obtained by the UDify tool using the multilingual model with Universal Dependency corpora as presented by Kondratyuk and Straka (2019).

[19] The UDify tool does not provide enhanced dependency graph analysis, thus, this column is annotated with an underscore in this study.

Three languages present LAS values lower than 80 (i.e. Irish, Lithuanian and Maltese). For 9 languages, the score is between 80 and 90, and for the rest (12 languages), the LAS is higher than 90. Thus, for the languages with the lowest scores, the bias regarding the dependency relation labels is the highest. The annotated texts are also publicly available[20] and can be used in a large variety of NLP studies requiring texts annotated following the Universal Dependencies framework.

3.4 Typological Approaches for Dependency Parsing Improvement

In this section, we describe briefly how typological approaches can be used for the improvement of dependency parsing scores, and we detail the results of specific corpus-based syntactic typological strategies using the Parallel Universal Dependencies collection.

3.4.1 Correlation Between Quantitative Typological Approaches and Dependency Parsing Improvement

In a detailed study concerning dependency parsing improvement using quantitative typological strategies and parallel corpora (i.e. PUD), Alves et al. (2023) tested four different corpus-based approaches:

- MarsaGram all properties
- MarsaGram linear properties
- Head and dependent relative order
- Verb and object relative order

Each method is fully described in the following sub-sections. For each strategy, first, Alves et al. (2023) generated the language vectors by extracting specific syntactic information from the PUD datasets. Then, dissimilarity matrices were calculated using Euclidean and cosine distances. Then, languages were combined in pairs (i.e. concatenation of training sets), and these bilingual corpora were used to train dependency parsing models using the UDify tool. The typological classifications are presented as dendrograms obtained from dissimilarity matrices using hclust(method="ward.D2") R function.

With the distance values from the dissimilarities matrices and the empirical LAS obtained from the bilingual training of the UDify tool, the authors calculated the correlation coefficients (Pearson's and Spearman's) between the language distances

[20] https://dataverse.harvard.edu/dataset.xhtml?persistentId=doi:10.7910/DVN/ULWS1K

obtained for each method to check which typological strategy is the most efficient for predicting the LAS improvements. Negative correlation results were expected, as the higher the distance between languages, the lower should be the LAS improvement.

These strategies were also compared to the typological classification obtained with the lang2vec tool (Littell et al. 2017). In this specific case, the PUD languages are represented as language vectors composed of 41 syntactic features with valid values (i.e. 0.0, 0.33, 0.66 and 1.0). The total number of syntactic features in this tool is 103, but only 41 have valid values for all PUD languages.

3.4.1.1 MarsaGram with All Properties

MarsaGram is a tool for exploring tree-banks. It extracts context-free grammars (CFG) from annotated datasets that can be used for statistical comparison between languages as proposed by Blache et al. (2016). The latest release of this software can be downloaded from the ORTOLANG platform of linguistic tools and resources.[21]

This software identifies four types of properties from the corpora:

- Precede or linear: It describes the relative position of two elements (A precedes B) inside a sub-tree governed by a specific head. Each element is described by its part-of-speech (POS) and dependency relation (deprel) in the syntactic tree. Although being part of the same sub-tree, elements A and B are not necessarily syntactically linked. An example of a sentence with this property is presented in Fig. 3.3).
- Require: This property describes the cases where the presence of an element A requires the existence of an element B inside the sub-tree.
- Unicity: an element A has this property if inside the sub-tree it occurs only once (i.e. no other element with the same part-of-speech and dependency label is attested).
- Exclude: In this case, the presence of element A excludes the occurrence of element B inside the sub-tree.

Of the four properties described above, only the linear one is directly linked to word-order patterns on the surface level of the sentence. In total, 158,755 patterns were extracted from the PUD corpora. The distribution in terms of types of property is presented in Table 3.3.

Each language vector regarding the MarsaGram all properties strategy is composed of these features associated with the value corresponding to its frequency of occurrence inside the corpus.

[21] https://www.ortolang.fr/market/tools/ortolang-000917

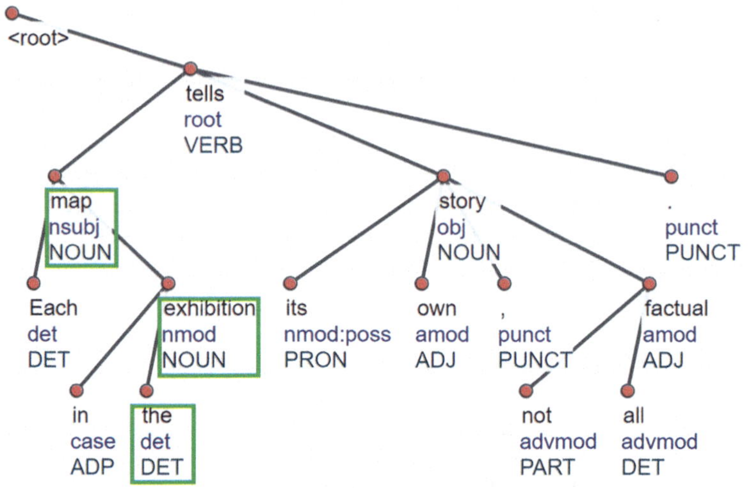

Fig. 3.3 Example of a sentence with the pattern NOUN_precede_DET-det_NOUN-nmod from the PUD English corpus. In the sentence "Each map in the exhibition tells its own story, not all factual.", the determiner ("the") 4 has the incoming relation det. It precedes the noun ("exhibition"), which has the incoming relation nmod. Both appear in the sub-tree headed by a noun ("map")

Table 3.3 Distribution of extracted features using MarsaGram in terms of properties

Property	Number of patterns	%
Linear	21,242	13.38
Require	6,189	3.90
Unicity	2,144	1.35
Exclude	129,180	81.37

3.4.1.2 MarsaGram with Linear Properties

As previously explained, the patterns with the linear property extracted with the MarsaGram tool are the ones that correspond to word-order phenomena inside sub-trees. Thus, Alves et al. (2023) decided to analyse them separately from patterns regarding other MarsaGram properties, especially because when all phenomena are considered, the large majority corresponds to the "exclude" property as presented in Table 3.3. By extracting just linear patterns from PUD corpora, language vectors were composed of 21,242 features.

3.4.1.3 Head and Dependent Relative Order

Beside the typological analysis provided by the MarsaGram tool, Alves et al. (2023) also proposed a quantitative approach concerning syntax, more specifically the head directionality parameter (i.e. whether the heads precede the dependents (right-branching) or follow them (left-branching) in the surface-level of the sentence) (Fábregas et al. 2015).

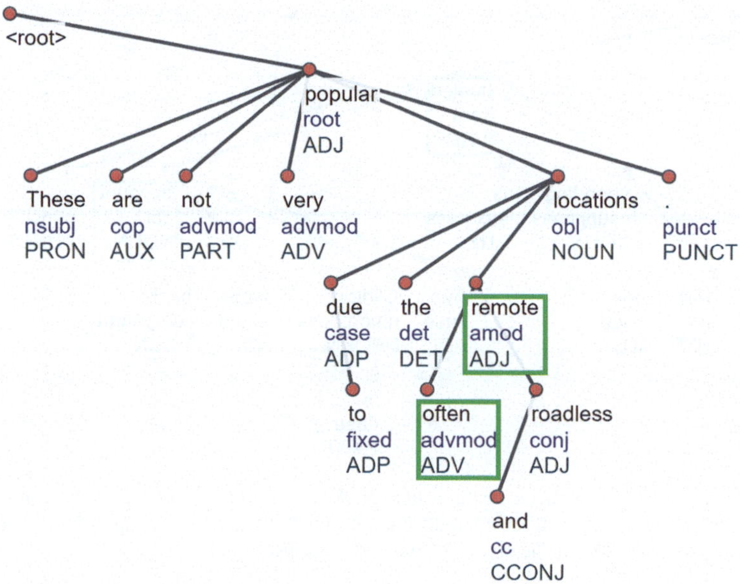

Fig. 3.4 Example of a sentence with two occurrences of the pattern ADV_advmod_precedes_ADJ. In the sentence "These are not very popular due to the often remote and roadless locations.", the adverb ("often") has the incoming relation advmod. It precedes its head, which is the adjective ("remote")

For this aim, the attested head and dependent relative position patterns (and their frequency) in the different PUD corpora were extracted using a Python script. All observed features extracted from the PUD corpora (2,890 in total) have been included in the language vectors. From this total, 1,374 features (47.5%) correspond to cases where the dependent precedes the head and 1,516 (52.5%) to right-branching patterns. In cases where a feature was not observed in a determined language, the value 0 was attributed to it.

Two examples of head and dependent relative position patterns are presented below:

- ADV_advmod_precedes_ADJ—head-final or left-branching—It means that the dependent, which is an adverb (ADV), precedes the head, which is an adjective (ADJ), and has the syntactic function of an adverbial modifier (advmod). The dependent can be in any position of the sentence previous to the head, not necessarily right before. An example of a sentence with this pattern is presented in Fig. 3.4.
- NOUN_obl_follows_VERB—head-initial or right-branching—In this case, the dependent (NOUN) comes after the head, which is a verb, and has the function of oblique nominal (obl). The dependent can be in any position after the head, not necessarily being right next to it. An example of a sentence representing this pattern is presented in Fig. 3.5.

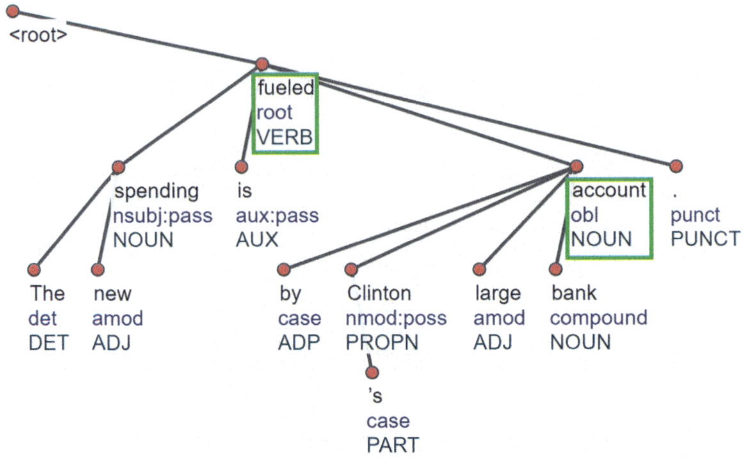

Fig. 3.5 Example of a sentence with the pattern NOUN_obl_follows_VERB. In the sentence "The new spending is fueled by Clinton's large bank account.", the noun ("account") has the incoming relation obl. It comes after its head, that is, the verb ("fueled")

This specific analysis of the head and dependent relative position corresponds to a quantitative interpretation of the head and dependent theory (Hawkins 1983), which considers that there is a tendency to organise heads and dependents in homogeneous word ordering. This author proposed a set of language types according to attested word-order phenomena concerning a limited list of elements as heads and dependents.

3.4.1.4 Verb and Object Relative Order

Inside the ensemble of features extracted for the analysis of the head and dependent relative position, it is possible to extract those regarding verbs and direct objects (deprel: "obj") for a specific analysis of these phenomena. Alves et al. (2023) decided to examine the position of these two elements in detail as they are key in typological studies such as the one proposed by Dryer (1992) where correlations are defined according to whether the verb comes before or after the object. In total, 13 OV (object preceding the verb) and 12 VO (verb preceding the object) features were attested in the PUD collection, allowing us to generate a 25-dimension language vector for each language.

3.4.1.5 Correlation Results

As previously described, Alves et al. (2023) calculated Pearson's and Spearman's correlation for each PUD language and for each typological strategy using the language distances from the dissimilarity matrices and the LAS deltas obtained when the languages were combined.

Table 3.4 Number of Pearson's correlations (moderate and strong) regarding all 20 PUD languages. The highest value regarding the total number is highlighted in bold. Euclidean distance, cos = cosine distance, Msg. all = Marsagram all features, Msg. lin. = Marsagram linear features, HD = Head and Dependent relative order, VO = Verb and Object relative order and L2V = lang2vec

	Msg. all Euc.	Msg. all cos	Msg. lin. Euc.	Msg. lin. cos	HD Euc.	HD cos	VO Euc.	VO cos	L2v Euc.	L2v cos
Strong	0	0	0	0	0	1	1	2	1	1
Moderate	3	8	3	10	7	7	5	2	6	5
Total	3	8	3	**10**	7	8	6	4	7	6

Table 3.5 Number of Spearman's correlations (moderate and strong) regarding all 20 PUD languages. The highest value regarding the total number is highlighted in bold. Msg. all = Marsagram all features, Euc. = Euclidean distance, cos = cosine distance, Msg. lin. = Marsagram linear features, HD = Head and Dependent relative order, VO = Verb and Object relative order and L2V = lang2vec

	Msg. all Euc.	Msg. all cos	Msg. lin. Euc.	Msg. lin. cos	HD Euc.	HD cos	VO Euc.	VO cos	L2v Euc.	L2v cos
Strong	0	1	0	0	1	2	2	0	1	1
Moderate	3	2	3	7	6	5	5	5	5	5
Total	3	3	3	**7**	**7**	**7**	**7**	5	6	6

When the obtained correlation value was between −0.7 and −0.5, it was considered a moderate inverse correlation and a strong one for values below −0.7. Tables 3.4 and 3.5 present the overall results concerning the number of cases (i.e. results for each PUD language) presenting either moderate or strong inverse correlation per typological strategy (Pearson's and Spearman's correlations respectively).

From the results displayed in Table 3.4, the typological approach that provides the language classification that correlates most with the empirical improvement in terms of LAS is the MarsaGram linear one concerning cosine distances. This approach presents a moderate or strong correlation for half of all PUD languages. It means that the linear order of components inside the same sub-tree seems a relevant factor that affects deep-learning systems.

The classic classification using lang2vec syntactic features only shows a strong or moderate correlation for 7 out of the 20 PUD languages. This score is even lower than other new methods such as Head and Dependent (cosine) and MarsaGram all properties (cosine).

Thus, as the MarsaGram linear strategy (cosine) is the one that is the most efficient for predicting dependency parsing improvements, we decided to compare the typological classification obtained via this method using parallel corpora to the same strategy applied to the automatically parsed text extracted from the Arquivo.pt 2019EUElections collection.

3.5 Corpus-Based Typological Approach Using Automatically Annotated Texts

In this section, we present the comparison between the typological classification obtained from the parallel corpora and the one regarding Arquivo.pt texts.

3.5.1 PUD MarsaGram with Linear (Cosine) Corpus-Based Typological Classification

Figure 3.6 presents the typological classification of ten European Union official languages (PUD corpora) generated with the MarsaGram linear (cosine) dissimilarity matrix provided by Alves et al. (2023).

As is apparent in Fig. 3.6, Romance languages are part of the same central cluster. The genealogical proximity between Portuguese and Spanish is also observed when the syntactic patterns (i.e. linear property) extracted using MarsaGram are considered. Moreover, some similarity between Czech and Polish (i.e. Slavic languages) is also noticeable in the dendrogram. On the other hand, Germanic languages present some particularities. While Swedish is positioned close to the Romance cluster, English is clustered with French, inside the Romance dendrogram group. German presents specific word order patterns (e.g. verb position) and is clustered with the Slavic languages in this analysis. Regarding Finnish, the only non-Indo-European language in our language set, it is isolated on the top-left side of the figure.

3.5.2 Arquivo.pt MarsaGram with Linear (Cosine) Corpus-Based Typological Classification

To avoid bias related to the size of corpora, we conducted the clustering analysis of the languages using automatically parsed texts extracted from the collection 2019EUElections, randomly selecting 1,000 sentences for each language (i.e. the same size as the PUD corpora).

Of these automatically annotated corpora, the MarsaGram tool identified 12,507 linear patterns that were used to form the language vectors associated with the frequency of occurrence in each corpus.

The obtained language vectors were used for the calculation of the cosine dissimilarity matrix. Then, the cluster analysis was conducted using these distance measures. The obtained dendrogram is presented in Fig. 3.7.

The corpus-based typological classification obtained with the patterns extracted from the Arquivo.pt automatically annotated corpora presents several similarities with the one built with data from the PUD collection.

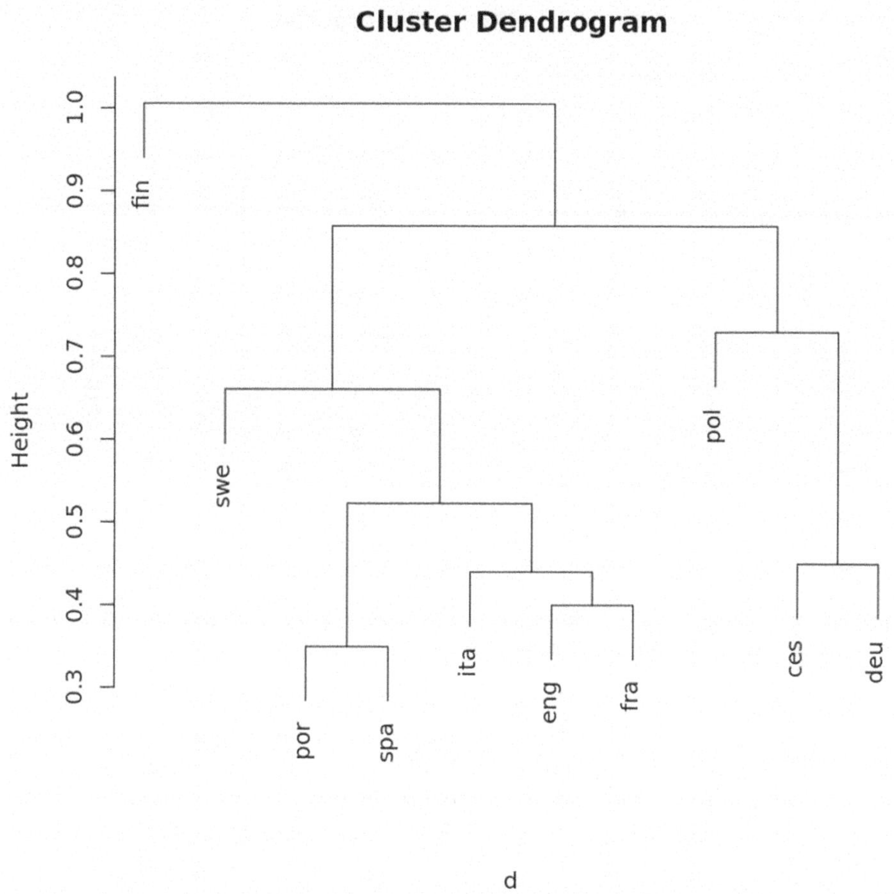

Fig. 3.6 Dendrogram with the classification of EU official languages using MarsaGram linear (cosine) method and PUD corpora

The Romance cluster is also easily identified in Fig. 3.7. While for PUD data, English was grouped inside this cluster, in this case, it forms a specific sub-group with Swedish. Another specificity of the 2019EUElections dendrogram is the well-defined Slavic cluster. Moreover, Finnish is also identified as an isolated language inside the language set, followed by German, which is, however, a bit closer to the main central cluster. However, when analysing in detail the Romance cluster, it is possible to notice that when the Arquivo.pt corpora are considered, Portuguese is clustered closer to French and not to Spanish as expected.

The differences between both dendrograms are most probably due to:

- the non-parallel nature of the Arquivo.pt sentences. Thus, corpora for some languages may contain more complex syntactic structures than for others.
- the automatic annotation which introduces errors regarding the syntactic analysis.

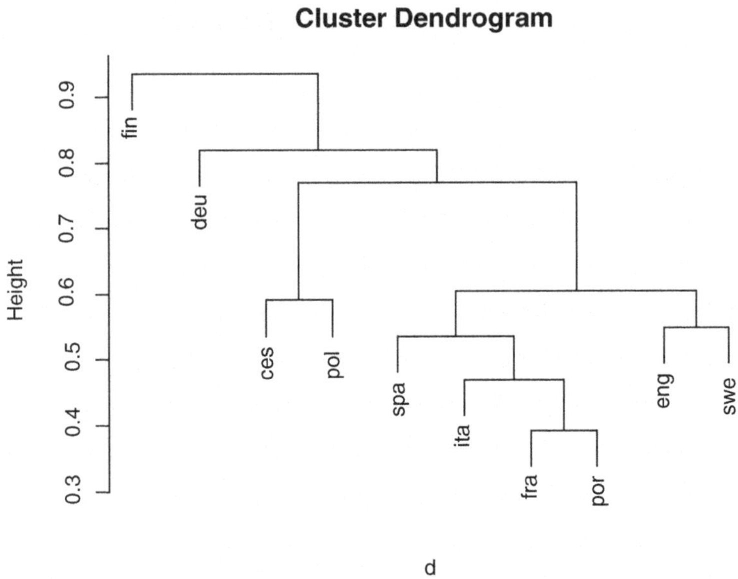

Fig. 3.7 Dendrogram with the classification of EU official languages using MarsaGram linear (cosine) method and the 2019EUElections corpora

Even though some differences can be observed, the usage of automatically annotated corpora seems useful in cases where non-parallel manually annotated corpora are available. The similarities between the two classifications show that the non-parallel corpora can provide some valuable information even though it contains some bias.

3.6 Conclusions and Future Work

This chapter presented an analysis of the robustness of a corpus-based typological strategy that is efficient for dependency parsing improvement. The idea was to compare the state-of-the-art corpus-based classification of ten European Union official languages developed with parallel corpora to the same method applied to non-parallel automatically annotated corpora.

First, we described the pipeline developed to extract and annotate texts from the Arquivo.pt 2019 European Parliamentary Elections collection (2019EUElections). This step generated two datasets composed of texts of all EU official languages: one composed of raw texts and the other also containing morphological and syntactic information obtained automatically. These datasets are available and can provide valuable linguistic information for other NLP studies.

Then, we presented how corpus-based typological approaches described in the literature can improve dependency parsing scores. Different methods have been established, but the language classification obtained via the comparison of language vectors composed with information regarding linear patterns extracted using the MarsaGram tool is the one with the highest correlation with the empirical parsing deltas.

Thus, using this aforementioned strategy, we compared the classification obtained using the Arquivo.pt data with the state-of-the-art one based on parallel corpora of ten EU official languages (PUD) via dendrograms. We showed that similarities can be observed; however, as the results are not totally similar, a more detailed analysis must be conducted by verifying the possible impact of these dissimilarities in experiments concerning language combination for dependency parsing improvement.

Acknowledgments The work presented in this chapter has received funding from the European Union's Horizon 2020 research and innovation program under the Marie Skłodowska-Curie grant agreement no. 812997 and under the name CLEOPATRA (Cross-lingual Event-centric Open Analytics Research Academy)

References

Alves D, Bekavac B, Zeman D, Tadić M (2023) Corpus-based syntactic typological methods for dependency parsing improvement. In: Proceedings of the 5th Workshop on Research in Computational Linguistic Typology and Multilingual NLP, Association for Computational Linguistics, Dubrovnik, pp 76–88. https://aclanthology.org/2023.sigtyp-1.8

Alzetta C, Dell'Orletta F, Montemagni S, Osenova P, Simov K, Venturi G (2020) Quantitative linguistic investigations across universal dependencies treebanks. In: Monti J, Dell'Orletta F, Tamburini F (eds) Proceedings of the Seventh Italian Conference on Computational Linguistics, CLiC-it 2020, Bologna, March 1–3, 2021. CEUR-WS.org, CEUR Workshop Proceedings, vol 2769

Ammar W, Mulcaire G, Ballesteros M, Dyer C, Smith NA (2016) Many languages, one parser. Trans Assoc Comput Linguist 4:431–444

Blache P, Rauzy S, de Montcheuil G (2016) Marsagram: an excursion in the forests of parsing trees. In: Proceedings of the Tenth International Conference on Language Resources and Evaluation LREC 2016, European Language Resources Association (ELRA). http://www.lrec-conf.org/proceedings/lrec2016/summaries/137.html

Dell'Orletta F, Venturi G, Montemagni S (2013) Linguistically–driven selection of correct arcs for dependency parsing. Computación y Sistemas 17. https://www.cys.cic.ipn.mx/ojs/index.php/CyS/article/view/1517

Dryer MS (1992) The greenbergian word order correlations. Language 68(1):81–138

Dryer MS, Haspelmath M (eds) (2013) WALS Online. Max Planck Institute for Evolutionary Anthropology, Leipzig. https://wals.info/

Fábregas A, Putnam M, Mateu J (2015) Contemporary linguistic parameters. Bloomsbury Publishing, London

Gerdes K, Kahane S, Chen X (2021) Typometrics: from implicational to quantitative universals in word order typology. Glossa: A J General Linguist 6(1):17

Gomes D (2022) Web archives as research infrastructure for digital societies: the case study of arquivo.pt. Archeion 2022(123):46–85. https://www.ejournals.eu/Archeion/2022/123/art/22601/

Hawkins JA (1983) Word Order Universals, vol 3. Elsevier, Amsterdam

Hawkins JA (2003) Efficiency and complexity in grammars: Three general principles. Nat Explanat Linguist Theory 121:152

Jurafsky D, Martin JH (2009) Speech and language processing: an introduction to natural language processing, computational linguistics, and speech recognition. Pearson Prentice Hall, Upper Saddle River

Kondratyuk D, Straka M (2019) 75 languages, 1 model: parsing universal dependencies universally. In: Proceedings of the 2019 Conference on Empirical Methods in Natural Language Processing and the 9th International Joint Conference on Natural Language Processing (EMNLP-IJCNLP), pp 2779–2795

Levshina N (2022) Corpus-based typology: applications, challenges and some solutions. Linguist Typol 26(1):129–160

Littell P, Mortensen DR, Lin K, Kairis K, Turner C, Levin L (2017) Uriel and lang2vec: representing languages as typological, geographical, and phylogenetic vectors. In: Proceedings of the 15th Conference of the European Chapter of the Association for Computational Linguistics: Volume 2, Short Papers, pp 8–14

Lynn T, Foster J, Dras M, Tounsi L (2014) Cross-lingual transfer parsing for low-resourced languages: an irish case study. In: Proceedings of the First Celtic Language Technology Workshop, pp 41–49

Östling R (2015) Word order typology through multilingual word alignment. In: Proceedings of the 53rd Annual Meeting of the Association for Computational Linguistics and the 7th International Joint Conference on Natural Language Processing (Volume 2: Short Papers), pp 205–211

Otter DW, Medina JR, Kalita JK (2018) A survey of the usages of deep learning in natural language processing. CoRR abs/1807.10854. http://arxiv.org/abs/1807.10854

Pires T, Schlinger E, Garrette D (2019) How multilingual is multilingual BERT? In: Proceedings of the 57th Annual Meeting of the Association for Computational Linguistics, Association for Computational Linguistics, Florence, pp 4996–5001. https://doi.org/10.18653/v1/P19-1493. URL https://aclanthology.org/P19-1493

Ponti EM, O'horan H, Berzak Y, Vulić I, Reichart R, Poibeau T, Shutova E, Korhonen A (2019) Modeling language variation and universals: a survey on typological linguistics for natural language processing. Comput Linguist 45(3):559–601

Sinnemäki K (2014) Complexity trade-offs: a case study. In: Measuring grammatical complexity. Oxford University Press, Oxford, pp 179–201

Straka M, Hajic J, Straková J (2016) Udpipe: trainable pipeline for processing conll-u files performing tokenization, morphological analysis, pos tagging and parsing. In: Proceedings of the Tenth International Conference on Language Resources and Evaluation (LREC'16), pp 4290–4297

Üstün A, Bisazza A, Bouma G, van Noord G (2020) Udapter: language adaptation for truly universal dependency parsing. arXiv preprint arXiv:200414327

Wälchli B (2009) Data reduction typology and the bimodal distribution bias. Linguist Typol 13:77–94

Zeman D, Hajic J, Popel M, Potthast M, Straka M, Ginter F, Nivre J, Petrov S (2018) Conll 2018 shared task: multilingual parsing from raw text to universal dependencies. In: Proceedings of the CoNLL 2018 Shared Task: Multilingual Parsing from Raw Text to Universal Dependencies, pp 1–21

Chapter 4
Processing Multimodal Information: Challenges and Solutions for Multimodal Sentiment Analysis and Hate Speech Detection

Sherzod Hakimov, Gullal S. Cheema, and Ralph Ewerth

Abstract This chapter explores the challenges and solutions for processing multimodal information, specifically in the context of multimodal sentiment analysis and hate speech detection. The increasing amount of multimodal data, such as text, images and videos, presents unique challenges for machine learning algorithms. These challenges include the integration and fusion of information from multiple modalities to acquire the overall context. In this chapter, first, we present an overview of recent developments on multimodal learning techniques in the context of sentiment and hate speech detection; second, we present a multimodal model that combines different visual aspects and features for multimodal sentiment detection; and third, we present a multi-task multimodal model for misogyny detection in multimodal memes.

4.1 Introduction

Sentiment analysis and hate speech detection have seen significant attention in recent years, with various model architectures developed to address these challenges. While existing methodologies have provided valuable insights, investigating multimodality remains a relatively unexplored area in these domains. This chapter

S. Hakimov (✉)
Computational Linguistics, University of Potsdam, Potsdam, Germany
e-mail: sherzod.hakimov@uni-potsdam.de

G. S. Cheema
L3S Research Center, Leibniz Universität Hannover, Hannover, Germany
e-mail: gullal.cheema@tib.eu

R. Ewerth
TIB – Leibniz Information Centre for Science and Technology, Hannover, Germany

L3S Research Center, Leibniz Universität Hannover, Hannover, Germany
e-mail: ralph.ewerth@tib.eu

I. Marenzi et al. (eds.), *Event Analytics across Languages and Communities*,
https://doi.org/10.1007/978-3-031-64451-1_4

aims to bridge the gap by presenting two model architectures tailored towards detecting sentiment or hate speech (specifically misogyny) in multimodal data (image-text pairs). For sentiment detection in multimodal data, we present a comprehensive analysis of multiple feature encoders into a single model. Another contribution is made by evaluating two well-known datasets transparently, which enables fair comparison for future work. The detection of misogynous memes is another challenge that is faced by many on social media, but without much prior work that can process such multimodal content. Our methodology uses pre-trained multimodal encoders and fuses them into a single joint embedding space. Both model architectures are fine-tuned on the respective data and compared against existing approaches. The presented solution provides a comprehensive framework for multimodal sentiment analysis, while the extension to misogynous meme detection demonstrates the adaptability and applicability of the proposed methodology across different domains. The experimental evaluation shows that both solutions have obtained the highest performance on datasets while revealing each building block's importance in a multimodal architecture.

Next, we present the overview of related work on multimodal learning and its extensions to sentiment and hate speech detection. We then present two separate approaches for building models to detect sentiment and hate speech in multimodal data and finally conclude the chapter with the findings.

4.2 Background and Related Work

This section briefly discusses developments in multimodal learning and abstract multimodal understanding tasks such as hate speech and sentiment detection. Following this, we review recent approaches, datasets and research progress in multimodal sentiment and hate speech detection.

4.2.1 Multimodal Learning

Multimodal learning is an exciting research field that has been established for quite some time. The approach involves the utilisation of two or more modalities, such as text, image and audio, to extract meaningful information that can be used to achieve specific objectives. This approach leverages the richness of diverse data types to learn better models. While it can encompass various combinations of modalities, our primary focus here is the fusion of vision and language. Recent years have witnessed a proliferation of deep learning techniques that have shifted the emphasis of multimodal research towards task-agnostic unsupervised representation learning across expansive multimodal datasets. One of the primary objectives of multimodal learning is to bridge the semantic gap, which is the discrepancy between low-level features and higher-level semantic interpretations (Smeulders et al. 2000). Bridging this gap involves effectively integrating contributions from disparate modalities to derive semantics that can pave the way for enhanced performance in prediction tasks like sentiment analysis in videos. This approach (Poria et al. 2017) often capitalises on visual and audio modalities to produce more precise results. Some noticeable areas and applications with image and text as modalities that have made significant progress in the last decade are image captioning (Karpathy and Fei-Fei 2015; Hossain et al. 2019), cross-modal retrieval (Socher et al. 2014; Zhen et al. 2019) and text-to-image generation (Mansimov et al. 2016; Qiao et al. 2019; Ramesh et al. 2021).

The fusion of vision and language, often referred to as vision-language (V-L) learning, has experienced rapid advancements, particularly due to deep learning innovations. Early models applied to V-L tasks employed architectures like recurrent neural networks (RNNs) (Rumelhart et al. 1986) and convolutional neural networks (CNNs) (LeCun et al. 1995) to individually process text or image data. With the introduction of the transformer model by Vaswani et al. (2017) and later models like BERT (Devlin et al. 2019) and GPT-3 (Brown et al. 2020), the paradigm of pre-training on large datasets and then fine-tuning for specific tasks has become popular in vision-language learning. There has been a surge in the development of vision-language pre-trained models that capitalise on large-scale image-text datasets to learn universal cross-modal representations. Recent research on fusion schemes, pre-training tasks, datasets and encoder architectures can be reviewed for further details in this survey (Du et al. 2022). In essence, the strength of V-L pre-trained models lies in their ability to amalgamate different single-modal encoders, intricate V-L interaction mechanisms and diverse pre-training strategies to achieve impressive results.

4.2.2 Sentiment

Sentiment detection has been extensively explored for textual social media data, with earlier lexicon-based approaches (like *SentiStrength* from Thelwall et al. 2010) evolving to statistical and machine learning-based classification in the last decade. The deep learning era later ushered in more intricate architectures like convolutional neural networks (CNNs) (Alayba et al. 2018; Kim 2014) and long short-term memory (LSTM) networks (Huang et al. 2016) optimised for tweet sentiment classification. In order to integrate traditional and modern techniques, Shin et al. (2017) combined lexicons with a CNN using an attention mechanism.

With platforms like Instagram, Flickr and Twitter becoming more visually driven, sentiment analysis has gained popularity due to the challenge of combining information from two or more modalities that influence sentiment. Early on, Cao et al. (2016) extracted low-level visual representations, *SentiBank* visual concepts and lexicon-based features from text to predict sentiment using late fusion strategies. To capture high-level concepts in both image and text, Cai and Xia (2015) as well as Yu et al. (2016) trained a shallow CNN for text and a deep CNN for images with shared representation to predict sentiment and achieved much better performance than the previous approaches. For the MVSA (Niu et al. 2016) dataset, in particular, several techniques improved the performance by focusing on cross-modal representations that capture the influence of modalities towards the sentiment. Xu and Mao (2017) proposed an approach using deep CNN with pre-trained features that encode object and scene information from images and aggregated it with contextual *GloVe* (Global Vectors for Word Representation) (Pennington et al. 2014) word embeddings from the text. They used visual feature-guided attention to capture the influence of visual features over word embeddings instead of simply concatenating them to predict the sentiment. Later, Xu et al. (2018) proposed a co-memory attention mechanism using similar features to capture the interaction between two modalities and their influence on the sentiment. Similarly, Jiang et al. (2020) proposed another attention mechanism where they used both cross-modal attention fusion followed by modality-specific CNN-gated feature extraction to learn a better representation. They used *ImageNet* (Russakovsky et al. 2015) pre-trained *ResNet* (He et al. 2016) for visual features and experimented with *GloVe* and *BERT* (Devlin et al. 2019) embeddings for textual features to achieve better results than previous approaches. Recently, Li et al. (2022) exploited contrastive learning and data augmentation on both text and image modalities, achieving performance boosts across various datasets, including MVSA.

4.2.3 Hate Speech

The detection of hateful content has been mostly studied from the textual perspective based on the computational linguistics and natural language processing (NLP) fields. However, hateful content on social media can be found in other forms, such as videos, a combination of text and images or emoticons. Misogynous content detection is especially challenging when textual and visual modalities are combined in a single context, e.g. an overlay text embedded on top of an image, also known as *meme*. Recent efforts in multimodal representation learning (Lu et al. 2019; Radford et al. 2021) pushed the boundaries of solving such problems by combining visual and textual representations of the given content. Several datasets have been proposed using multimodal data (Gomez et al. 2020; Kiela et al. 2020b; Sharma et al. 2020; Pramanick et al. 2021; Suryawanshi et al. 2020; Menini et al. 2020) for various tasks related to hate speech. Each dataset includes an image and corresponding text, which is either an overlay text embedded on an image or a separate accompanying text such as tweet text. Existing model architectures that are evaluated on such benchmark datasets use a combination of textual and visual features extracted from pre-trained models (Kiela et al. 2020a).

4.3 Multimodal Sentiment Analysis

Social media has become a phenomenon in terms of its usage by the general public, traditional media and enterprises and as a forum for discussing research in academia. With the evolution of the Internet, social media sites, in particular, have become multimodal in nature with content including text, audio, images and videos to engage different senses of a user. Similarly, sentiment analysis techniques have also progressed from extensively explored text-based (Liu and Zhang 2012; Medhat et al. 2014) to multimodal sentiment analysis (Soleymani et al. 2017) of image-text pairs or videos. With two or more modalities, the problem becomes more challenging since every modality might influence the overall sentiment differently, and modalities can have a complex interplay. For image-text pairs, this is even harder as images are perceived as a whole, whereas text is read sequentially. Existing approaches focus on different types of features (Niu et al. 2016; Xu and Mao 2017) and complex attention mechanisms (Jiang et al. 2020; Xu et al. 2018) to capture the inter-dependencies between image and text to build multimodal models.

Psychological studies have found that human visual attention generally prioritises emotional content over non-emotional content (Brosch et al. 2010; Compton 2003). A recent study by Fan et al. (2018) evaluated the inter-relationships of image sentiment and visual saliency in deep convolutional neural network (CNN) models. They proposed a model that prioritises emotional objects over other objects to predict sentiment, just like human perception. It indicates that to learn a multimodal model for sentiment prediction, visual features should contribute and consider different objects, facial expressions and other salient regions in the image. Besides, to learn a multimodal model for sentiment detection, extracted features from two modalities need to be combined in a way that reflects the overall sentiment of the image-text pair. Even though the existing approaches (Xu and Mao 2017; Xu et al. 2018; Jiang et al. 2020) for image-text sentiment detection proposed complex attention mechanisms over different types of features, they fall short on the analysis of features and the number of visual features used and lack a reproducible evaluation, which hampers progress in this field.

In this section, we study the impact of different visual features in combination with contextual text representations for multimodal tweet sentiment classification and present a comprehensive comparison with six state-of-the-art methods (Cheema et al. 2021). In contrast to previous work, we investigate four different visual feature types: facial expression, object, scene and affective image content. We utilise a simple and efficient multimodal neural network model (Se-MLNN, Sentiment Multi-Layer Neural Network) that combines several visual features with contextual text features to predict the overall sentiment accurately. In our experiments, we also test the recently proposed *CLIP* model (contrastive Language-Image Pretraining Radford et al. 2021), which is specifically trained on millions of image-text pairs and reports impressive zero-shot performance on image classification datasets like *ImageNet* (Russakovsky et al. 2015) and *Places365* (Zhou et al. 2018). We use this model instead of pre-trained multimodal transformers due to the volume and variety of data it exploited for pre-training, which makes it attractive for different kinds of visual recognition tasks.

We use the publicly available benchmark MVSA (Niu et al. 2016) (Multi-View Social Data) that consists of two different datasets of tweets and corresponding images. We provide a detailed analysis of image and text features complemented with an extensive experimental study and outline the limitations of existing approaches. All existing approaches for the MVSA datasets use randomly generated, unpublished train and test splits, making it impossible to reproduce results or fairly compare them. Thus, we apply k-fold cross-validation so that every dataset sample is tested once. We share the source code and the new dataset splits used in this chapter.[1]

The main idea is to exploit and investigate different kinds of high-level visual features and combine them with a textual model. The use of channel features as a sequence in conjunction with word embeddings by previous approaches limits their model to two modalities or types of features (Jiang et al. 2020; Xu and Mao 2017; Xu et al. 2018). In contrast, we aim to investigate the impact of our suggested high-level visual features, which are objects, scenes or places, facial expressions and the overall affective image content in a more efficient and less complex framework.

A crucial difference between our approach and recent multimodal tweet sentiment approaches (Jiang et al. 2020; Xu et al. 2018) is that we use pooling to get one embedding per image instead of using channel features from a pre-trained CNN as a sequence. Similarly, we use a pooling strategy for getting one embedding per tweet from a textual model. Another difference is that instead of relying on learning bi-attention weights from a limited amount of data, we use a multi-layer neural network to combine different features to influence the sentiment. To investigate the impact of our suggested high-level visual features, we propose a three-layer neural network, where the first two layers aggregate features from different modalities and the third layer is used for the classification of sentiment. The architecture of our approach is shown in Fig. 4.1. The training details are provided in Sect. 4.3.5. Next, we describe the individual models for each modality and their encoding process to understand the proposed approach.

4.3.1 Visual Features

This section describes and presents various visual features used in the proposed approach.

[1] https://github.com/cleopatra-itn/fair_multimodal_sentiment

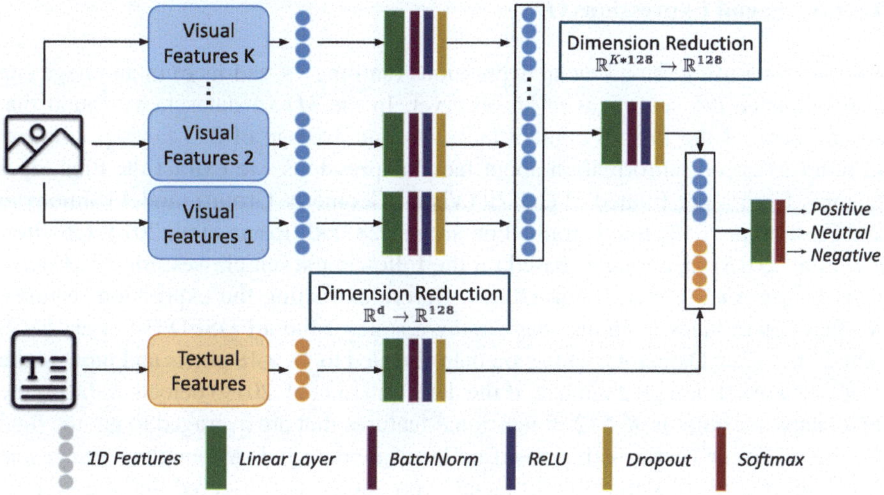

Fig. 4.1 Se-MLNN: Proposed architecture for multimodal sentiment classification. Here d in $\mathbb{R}^d$ is different for every feature and is provided in Sects. 4.3.1, 4.3.2 and 4.3.3. Every feature irrespective of the dimension d is projected down to 128 (First Layer) in order to keep the number of parameters low and not to introduce feature bias. The final layer ($\mathbb{R}^{256} \rightarrow \mathbb{R}^3$) is followed by Softmax that outputs the probability of each sentiment. This figure has been designed using images from Flaticon.com

4.3.1.1 ImageNet Features (E_o)

Different objects in a picture can incite a particular sentiment in a person. For instance, a cute dog or flowers might bring a positive sentiment, whereas a snake may incite a negative sentiment depending on the context. To encode objects and the overall image content, we extract features from a pre-trained *ResNet* model (He et al. 2016) trained on *ImageNet* (Russakovsky et al. 2015). We use *ResNet*-50 and its last convolution layer to extract features instead of the object categories (final layer). The final convolutional layer outputs 2,048 feature maps each of size 7×7, which is then pooled with a global average to get a 2048-dimensional vector.

4.3.1.2 Place and Scene Features (E_s)

A scene or a place can also incite different sentiments in a person. For instance, a candy store might bring a positive sentiment, whereas a catacomb might incite a negative sentiment depending on the context. To encode the scene information of an image, we extract features from a pre-trained *ResNet* (He et al. 2016) model trained on *Places365* (Zhou et al. 2018). In this case, we use *ResNet*-101 and follow the same encoding process as above.

4.3.1.3 Facial Expressions (E_f)

The presence of faces and facial expressions (smiling vs. sad face) in an image can also influence the sentiment in an observer. In the *MVSA* dataset, we found that around 50% of the images contain faces with an average of 2–3 faces per image. In order to encode information about facial expressions, we extract the final layer features from a pre-trained[2] *VGG-19* (Visual Geometry Group) model (Simonyan and Zisserman 2015) that is trained on around 28.000 (Erhan et al. 2013; Goodfellow et al. 2015) face images based on the following seven classes: *angry*, *disgust*, *fear*, *happy*, *sad*, *surprise* and *neutral*. Before extracting the expression features, we first detect faces from an image using a state-of-the-art DSFD (Li et al. 2019) (Dual Shot Face Detector), which are then rescaled to 48×48 pixels and input to the *VGG* network. For a given image, if the detector (Li et al. 2019) detects K faces, the *VGG* network outputs K 512-dimensional features that are averaged to get the final feature vector or vector with zeros if no faces are detected. Although an image can have multiple faces with different facial expressions, we observed that the average of feature embeddings is the best aggregation method compared to other pooling strategies, like averaging the facial expression predictions and maximum probability feature.

4.3.1.4 Affective Image Content (E_a)

Overall affective image content can also be important for multimodal sentiment detection, and research in this area has made rapid progress in recent years with famous datasets from popular social media image-sharing Web sites such as Flickr and Instagram. To encode the overall emotion, we first fine-tune a *ResNet*-50 *ImageNet* model on publicly available FI (Flickr & Instagram) dataset (You et al. 2016) and extract the last layer convolution features as described above for object and scene embeddings. The dataset consists of around 23,000 training images and 8 emotion classes: *amusement*, *anger*, *awe*, *contentment*, *disgust*, *excitement*, *fear* and *sadness*.

4.3.2 *Textual Features*

Since the context and meaning of the words are equally important for the influence of the whole sentence towards the sentiment, we use *RoBERTa*-Base (Liu et al. 2019) (Robustly optimised BERT approach) to extract contextual word embeddings and employ different pooling strategies to get a single embedding for the tweet.

[2] https://github.com/WuJie1010/Facial-Expression-Recognition.Pytorch

We experimentally found that the average of the last four layers is the most useful, and we use this embedding for all our *RoBERTa* embedding experiments. We finally take an average over the word embeddings to get the single text embedding of 768 dimensions for every tweet. For pre-processing text, we normalise text following Baziotis et al. (2017) and conduct three experiments by keeping (E_T^{+HT}) and removing (E_T^{-HT}) the hashtags from the text and also on the raw tweet (E_T^{RAW}) text.

4.3.3 Multimodal Features

We use the recently proposed multimodal model *CLIP* (Radford et al. 2021) that is trained on 400 million image-text pairs collected from the Internet. The model is trained to predict which caption goes with which image, and in doing so, it learns expressive image representation without the need for millions of labelled training examples. In comparison to multimodal transformers (Lu et al. 2019; Li et al. 2020), the model uses pairwise learning over n-pairs of image and text and does not use any cross-attention mechanism to learn multimodal features. This makes the model easy to use, as image and text embeddings can be independently computed from the respective image and text encoders. Because of the variety and large amount of data, the model shows competitive zero-shot recognition performance on 30 different computer vision datasets compared to their supervised baselines. This suggests that the amount and quality of visual information encoded in the visual features of the model are much better than the *ImageNet* and *Places365* supervised pre-trained models.

We use a publicly available *CLIP* model[3] and a variant that has a visual and textual transformer as image and text encoder backbones. We extract image and text features from the model where image (C_I) and text embeddings are 512-dimensional vectors. For text, we use the same pre-processing as used for textual models, and this results in three types of text embeddings, with hashtags (C_T^{+HT}), without hashtags (C_T^{-HT}) and raw text (C_T^{RAW}).

4.3.4 Dataset and Training Details

In this section, we present the datasets used for training and evaluating models, training and hyperparameter details and baseline models used for comparison with the proposed approach.

[3] https://github.com/openai/CLIP

4.3.4.1 Datasets

We use the *MVSA-Single* (MVSA-S) and *MVSA-Multiple* (MVSA-M) datasets (Niu et al. 2016) to test our model. The two datasets contain 4,869 and 19,598 image-text pairs from Twitter, where both image and text are annotated with a separate label by a single annotator (*MVSA-S*) or three annotators (*MVSA-M* with three annotations for each sample), respectively.

For a fair comparison and to be consistent with previous work, we process to get the multimodal label and filter the two datasets according to Xu and Mao (2017); Xu et al. (2018), which results in 4,511 and 17,025 image-text pairs, respectively. To summarise, the majority of the assigned class labels over each pair is the pooled label, and the final label falls under three cases: (1) label is valid and same if both have the same label, (2) label is valid and a polar label if either is positive or negative and the other is neutral and (3) tweet is a conflict and filtered if image and text have opposite polarity labels. The *MVSA-Single* dataset consists of 470 *neutral*, 2,683 *positive* and 1,358 *negative* samples, while the *MVSA-Multiple* dataset consists of 4,408 *neutral*, 11,318 *positive* and 1,299 *negative* samples.

4.3.4.2 Evaluation and Comparison

We conduct tenfold cross-validation where every split ends up with 8:1:1 ratio data and the same label distribution in training, validation and test set. For our ablation study, we report the average accuracy and weighted F1 scores (according to class size) over the tenfold in Tables 4.1 and 4.2. For comparison with two other approaches, we report minimum, maximum and average accuracy and weighted-F1 measure for all our runs. Our results cannot be directly compared with some previous approaches' reported results (taken from Jiang et al. 2020) and are only here for reference (marked with †). We evaluate the publicly available *MultiSentNet* (Xu and Mao 2017) implementation[4] and re-implemented the *FENet* (Jiang et al. 2020) model to compare their results (marked with *) in Table 4.3.

4.3.4.3 Training Details

We use cross-entropy as an objective function and the Adam (adaptive moment estimation) (Kingma and Ba 2015a) optimiser for updating the neural network parameters. We observe that the *MVSA-M* dataset label pooling strategy results in noisy labels, and to mitigate that, we use label smoothing so that the model does not become overconfident. With a smoothing factor of $\alpha = 0.1$ and $K = 3$ classes, the

[4] https://github.com/xunan0812/MultiSentiNet

Table 4.1 Unimodal and multiple visual feature results for *MVSA-Single* and *MVSA-Multiple*. Accuracy and F1 scores are averaged over tenfold

Features	MVSA-single		MVSA-multiple	
	ACC	F1	ACC	F1
E_T^{RAW}	68.46	66.01	62.83	57.70
E_T^{+HT}	68.88	66.49	60.09	55.58
E_T^{-HT}	67.50	64.50	65.65	**59.61**
C_T^{+HT}	**71.00**	**68.47**	58.52	54.50
C_T^{-HT}	68.72	65.59	**65.70**	59.43
E_o	64.69	61.40	65.28	56.63
E_s	64.66	61.51	65.13	56.00
E_a	64.89	61.65	64.85	56.02
E_f	59.63	48.48	**66.41**	53.21
C_I	**72.09**	**70.03**	65.42	**59.22**
$C_I + E_o + E_s + E_f$	**70.29**	**69.51**	63.65	59.87
$C_I + E_a + C_f$	69.70	68.66	**63.79**	**60.33**
$E_o + E_s + E_a + E_f$	66.42	65.40	63.49	58.58
$E_o + E_a + E_f$	66.10	65.05	63.75	58.89

Table 4.2 Multimodal results for *MVSA-Single* and *MVSA-Multiple*. Accuracy and F1 scores are averaged over tenfold

Features	MVSA-single		MVSA-multiple	
	ACC	F1	ACC	F1
	E_T^{+HT}		E_T^{-HT}	
E_o	71.80	70.09	66.28	60.98
E_s	72.53	70.77	65.80	60.59
E_a	71.98	70.20	66.27	61.13
$E_o + E_f$	72.85	71.57	66.01	62.51
$E_s + E_a$	72.80	71.32	66.12	61.49
$E_o + E_a + E_f$	72.93	71.80	66.19	62.57
$E_s + E_a + E_f$	72.93	71.69	66.31	**62.76**
C_I	**75.33**	73.76	**66.35**	61.89
$C_I + E_f$	75.00	**73.96**	66.08	62.52
$C_I + E_a$	73.95	72.86	65.18	62.03
$C_I + E_s + E_f$	74.73	73.60	66.02	62.51
	C_T^{+HT}		C_T^{-HT}	
C_I	**74.97**	**73.32**	**66.09**	61.27
$C_I + E_f$	74.00	72.58	65.32	**61.34**
$C_I + E_a$	73.29	72.15	64.68	61.13
$C_I + E_s + E_f$	73.89	72.68	65.43	61.49

Table 4.3 Comparison results for *MVSA-Single* and *MVSA-Multiple*. Results marked with † are taken from Jiang et al. (2020), and * are results of re-implemented models

	MVSA-single			MVSA-multiple		
Baseline	ACC		F1	ACC		F1
SentiBank and SentiStrength† (Borth et al. 2013)	52.05		50.08	65.62		55.36
CNN-Multi† (Cai and Xia 2015)	61.20		58.37	66.30		64.19
DNN-LR† (Yu et al. 2016)	61.42		61.03	67.86		66.33
MultiSentiNet† (Xu and Mao 2017)	69.84		69.63	68.86		68.11
CoMN(6)† (Xu et al. 2018)	70.51		70.01	70.57		70.38
FENet-BERT† (Jiang et al. 2020)	74.21		74.06	71.46		71.21
	ACC			F1		
	Avg	Min	Max	Avg	Min	Max
Models	MVSA-single					
MultiSentiNet*	63.27	57.87	69.25	59.12	57.83	63.61
FENet-BERT*	69.02	63.76	71.67	67.30	61.42	69.97
Se-MLNN($E_o + E_T^{+HT}$)	71.80	66.96	76.72	70.09	66.17	74.46
Se-MLNN($C_I + E_T^{+HT}$)	**75.33**	**70.51**	**82.04**	**73.76**	**69.82**	**81.14**
	MVSA-multiple					
MultiSentiNet*	63.08	54.32	67.10	59.12	54.43	58.57
FENet-BERT*	**68.61**	**61.47**	**74.40**	**65.80**	**60.84**	**73.56**
Se-MLNN($E_o + E_T^{-HT}$)	66.28	58.60	69.51	60.98	54.15	64.39
Se-MLNN($C_I + E_T^{-HT}$)	66.35	59.54	70.27	61.89	55.43	65.33

one-hot encoded label vector y becomes:

$$y_{ls} = (1 - \alpha) * y + \alpha/(K - 1) \tag{4.1}$$

The learning rate is set to 2×10^{-5}, and all the models are trained for 100 epochs. We decay the learning rate by a factor of 10 if the validation loss does not decrease for five epochs. A batch size of 32 and 128 is used for *MVSA-S* and *MVSA-M*, respectively. To avoid overfitting, a dropout with the ratio of 0.5 is applied after all the intermediate linear layers. We save the best model (of all epochs) according to the lowest validation loss while training and use it for testing. We use PyTorch[5] for our experiments and extract ImageNet features from its publicly available *ImageNet* pre-trained *ResNet-50* (He et al. 2016) model. We train a *ResNet-101* model on the *Places365* dataset and use it for extracting scene features.

[5] https://github.com/pytorch/pytorch

4.3.5 Results

As listing all feature combinations is ineffective, we report the most informative ones and reflect the use of each feature type. Also, we only show a maximum combination with four types of features (visual + textual), as no considerable improvement was observed with five features.

4.3.5.1 Unimodal Results

The unimodal experimental results reflect the use of either different textual or visual features discussed in Sect. 4.3. The purpose is to get stronger unimodal baselines with a simple neural network. Table 4.1 presents the evaluation of unimodal textual and visual features for the two datasets. For *MVSA-S*, we found out that including hashtag words in the text (E_T^{+HT}/C_T^{+HT}) gives slightly better performance than removing hashtags or using raw tweet (E_T^{RAW}) text. On the other hand, excluding hashtag words (E_T^{-HT}/C_T^{-HT}) from the text works better for the larger *MVSA-M*, where the inclusion of hashtags degrades the average accuracy by almost 6%. The E_T^{RAW} performance is slightly better than E_T^{+HT} for *MVSA-M* as *RoBERTa's* (Liu et al. 2019) word piece tokenisation tokenises hashtags differently than the pre-processing we used in E_T^{+HT}. This also shows that pre-processing noisy tweet text can be crucial for tasks like sentiment detection. For single visual-only models, we can see that all the visual features except facial expressions (E_f) are helpful for sentiment detection. For both modalities, *CLIP* features (C_I and C_T) outperform all the other features by 2–6% for *MVSA-S* and show similar or slightly better results for *MVSA-M*. This suggests that the pre-training strategy used in *CLIP* learns expressive visual and textual features, which can be used in multimodal downstream tasks. Interestingly, C_I outperforms all other unimodal features for both datasets.

4.3.5.2 Visual Combination Results

For the set of multiple visual features, we systematically combined them and observed improvements with three or more visual features. Due to space limitations, we only show some of them in Table 4.1 to highlight some aspects of the combinations that are discussed further. With visual features other than C_I, we see that the addition of each type of feature (like $E_o + E_a + E_f$) increases the performance in both datasets, especially the increase in accuracy and F1 measure by 1–4% on *MVSA-S*. Adding any other feature to C_I degrades the performance, indicating that other visual features are incompatible with C_I in our model, although they slightly increase the F1 score for *MVSA-M*. The improvement can be attributed to emotion features E_a and E_f, which shows that the affective image content is equally important in addition to object and scene information. Facial features alone perform the worst (very low F1) across datasets as almost 50% of the images have no detectable faces. For *MVSA-M* in particular, the combination of visual modalities only increases the F1 score and needs further analysis. On further inspection of the datasets, we ran OCR (optical character recognition) and found that around 15 and 24% of the images in *MVSA-S* and *MVSA-M*, respectively, have a considerable amount of text (>4 words). This could explain the limited to no performance improvement of combining visual features for the *MVSA-S* and *MVSA-M* datasets and the need to incorporate overlay and scene text recognition in future work.

4.3.5.3 Multimodal Results and Comparison

For multimodal experiments, we also systematically combine features by adding one type of visual feature to the model, some of which can be seen in Table 4.2. It can be observed that the combination of two modalities increases the sentiment prediction performance across the splits and considerably increases (by 4%) both measures on *MVSA-S* as shown ($C_I + E_T^{+HT}$) in Table 4.2. This improvement can be seen across all the splits for average, minimum and maximum values in Table 4.3, showing that the addition of modalities increases the best score and works for most of the splits. For the *MVSA-M*, the improvement is minimal with 1% accuracy and 3% F1 from unimodal models. Interestingly, visual features combined with *RoBERTa* (E_T^{+HT}/E_T^{-HT}) pooled features always outperform combinations with *CLIP's* text features (C_T^{+HT}/C_T^{-HT}) as some are shown in two separate grouped blocks in Table 4.2. This limited performance can be attributed to three issues: (1) considerably higher number of neutral samples that have a higher chance of getting classified as negative or positive; (2) the label pooling strategy (Xu and Mao 2017) used to pool labels from three annotators, which gives preference to positive or negative over the neutral label (see above) and results in a larger number of disputable labels; and (3) the model's inability to capture the interactions and differentiate between neutral and polar samples. When combining only two types of features for a fair comparison, our method ($E_o + E_T^{+HT}$) achieves better performance than both cross-attention based approaches for *MVSA-S*. This shows that a detailed cross-validation study is important to understand the effectiveness and limitations of a method. For *MVSA-M*, *FENet* (Jiang et al. 2020) achieves the best performance, which can be attributed to their use of self- and cross-attention to capture complex interactions between image and text. However, their reported results on a random split are different from our implementation's results. In the next section, we conduct an error analysis of misclassified samples and group these errors for further consideration. Also, combining all four visual modalities with text features did not improve the performance, possibly due to the increase in network parameters. We direct the reader to Cheema et al. (2021) for a comprehensive error analysis and discussion about limitations.

4.4 Multimodal Hate Speech Detection

Detection of hate speech has become a fundamental problem for many social media platforms such as Twitter, Facebook and Instagram. There have been many efforts by the research community and companies to identify the applicability of developed solutions. In general, hate speech is defined as *a hateful language targeted at a group or individuals based on specific characteristics such as religion, ethnicity, origin, sexual orientation, gender, physical appearance, disability or disease*. The hatred or contempt expressed towards women has been drastically increasing, as reported by Plan International (2020) and Vogels (2021). Detection of such misogynous content requires large-scale automatic solutions (Gasparini et al. 2018; Suryawanshi et al. 2020; Menini et al. 2020) and comprehensive annotation processes such as those defined by Zeinert et al. (2021).

The *SemEval-2022 Task 5: MAMI—Multimedia Automatic Misogyny Identification* (Fersini et al. 2022)[6]—is a new challenge dataset that focuses on identifying misogynous memes. The memes in this dataset are composed of an image with an overlay text. Some samples from the dataset with their corresponding class labels are provided in Fig. 4.2. The dataset includes two sub-tasks, as described below:

- Task-A: a basic task about misogynous meme identification, where a meme should be categorised either as misogynous or not misogynous
- Task-B: an advanced task where the type of misogyny should be recognised among potential overlapping categories such as stereotype, shaming, objectification and violence.

This section presents our participating model architecture under the team name *TIB-VA*. The model architecture is based on a neural model that uses pre-trained multimodal features to encode visual and textual content and combines them with an LSTM (long-short term memory) layer. Our proposed solution obtained the best result (together with two other teams) on the *Task-B*.

Our model architecture is a neural model that uses a pre-trained CLIP (Radford et al. 2021) model to extract textual and visual feature representations. We used the recently available *ViT-L/14* variant of CLIP. The tokens in the overlay text and the image are fed into *CLIP Text Encoder* and *CLIP Image Encoder*, respectively. The text encoder outputs a sequence of 768-dimensional vectors for each input token. These token vectors are then fed into an LSTM layer with a size of 256. The output from the image encoder is fed into a fully connected layer with a size of 256. The output from an LSTM layer for text and output from the fully connected layer for the image are fed into separate dropout layers (dropout rate of 0.2), concatenated and then fed into another fully connected layer with a size of 256. The final vector representation is then fed into separate sigmoid functions for each task. For Task-A, the sigmoid outputs a single value indicating misogyny probability. For Task-B,

[6] https://competitions.codalab.org/competitions/34175

Fig. 4.2 Four data samples from *MAMI—Multimedia Automatic Misogyny Identification*—with their corresponding class labels. Misogynous samples have additional sub-classes from *stereotype*, *shaming*, *objectification* and *violence*

each sub-class of misogyny (stereotype, shaming, violence, objectification) has a separate sigmoid function that outputs a probability value for the corresponding class. The model architecture is shown in Fig. 4.3. The source code of the described model is shared publicly with the community.[7]

[7] https://github.com/TIBHannover/multimodal-misogyny-detection-mami-2022

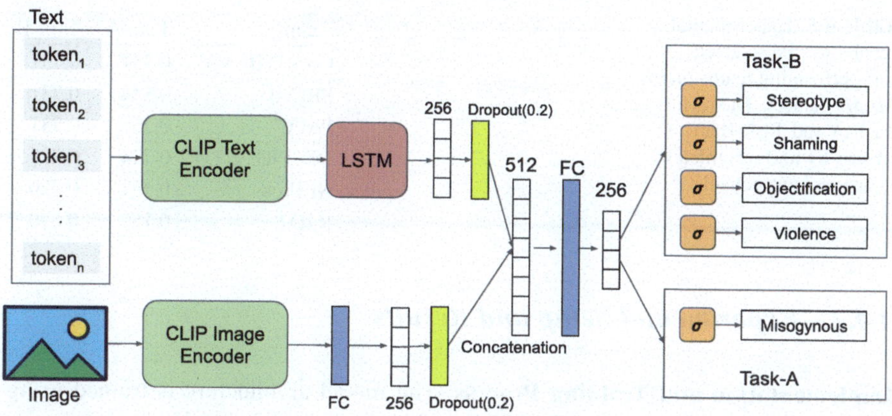

Fig. 4.3 The model architecture that combines textual and visual features to output probabilities for Task-A (misogynous) and Task-B (stereotype, shaming, objectification, violence). FC: Fully connected layer, σ: sigmoid function. Image taken from Hakimov et al. (2022)

4.4.1 Dataset

The *SemEval-2022 Task 5: MAMI—Multimedia Automatic Misogyny Identification* (Fersini et al. 2022)—aims at identifying misogynous memes by taking into account both textual and visual content. Samples from the dataset are given in Fig. 4.2. The dataset includes the overlay text extracted from an image. The challenge is composed of two sub-tasks. Task-A is about predicting whether a given meme is misogynous or not. Task-B requires models to identify sub-classes of misogyny (stereotype, shaming, violence, objectification) in cases where a given meme is misogynous. Task-B samples can have multiple labels, and a meme can have a single or all of the above sub-classes of misogyny. The train and test splits have 10,000 and 1,000 samples, respectively. The distribution of samples for the corresponding two sub-tasks are given in Table 4.4.

Table 4.4 Distribution of samples in Task-A and Task-B for train and test splits in the *MAMI—Multimedia Automatic Misogyny Identification* dataset

	Task-A		Task-B				
Splits	Misogynous	NOT	Shaming	Objectification	Violence	Stereotype	Total
Train	5,000	5,000	1,274	2,202	953	2,810	10,000
Test	500	500	146	348	153	350	1,000

Table 4.5 Experimental results for the selected top-performing teams on the *MAMI* dataset. The results on Task-A and Task-B are Macro-F1 and weighted F1 measures, respectively

Team	Task-A	Task-B
Ours (TIB-VA)	0.734	0.731
SRC-B	0.834	0.731
PAFC	0.755	0.731
DD-TIG	0.794	0.728
NLPros	0.771	0.720
R2D2	0.757	0.690

4.4.2 *Experimental Setup and Results*

Implementation and Training Process The model architecture is trained using Adam optimiser (Kingma and Ba 2015b) with a learning rate of *1e–4*, a batch size of *64* for a maximum of *20* epochs. We decrease the learning by half after every five epochs. We use 10% of the training split for validation to find the optimal hyper-parameters. The model architecture is implemented in Python using the PyTorch library.

Comparison of Top-Ranking Models The official evaluation results[8] for the top-performing teams are presented in Table 4.5. The results on Task-A and Task-B are macro-averaged F1 and weighted-averaged F1 measures, respectively. Our model architecture (team *TIB-VA*) achieves the best result (0.731) on Task-B along with the other two teams: *SRC-B* and *PAFC*. The *SRC-B* team has the highest performance on Task-A. We have also experimented with various model architectures, such as using Support Vector Machines (SVM), XGBoost and fully connected neural networks on the features extracted from vision and text encoders. However, none of these model architectures exceeded the performance of the one presented in Fig. 4.3. We have also experimented with using other vision encoders, such as ResNet models, but none of them resulted in a better performance when compared to the CLIP vision encoder.

4.5 Discussion

We have presented approaches that utilise multimodal features for building models to detect sentiment or hate speech in image-text pairs (see Sects. 4.3 and 4.4). Both approaches rely on pre-trained vision or text encoders that are then fused to form the multimodal content in the embedding space. The vision encoder component of CLIP has been shown to obtain the best performance when combined with some text encoder for both tasks, which can be explained by the amount of data the CLIP model has been trained on in comparison to vanilla ResNet models. Another

[8] https://competitions.codalab.org/competitions/34175#results

common pattern among both architectures is the use of a fusion component that merges dense vector representations from both modalities into one. It is an essential part of any model architecture trained on inputs from multiple modalities.

As mentioned before, both approaches rely on pre-trained text encoders. In the case of the hate speech detection model, the text and vision encoders are jointly pre-trained, whereas the model architecture for the sentiment detection learns to align the embeddings from both modalities by fine-tuning on the task data. One limitation of such approaches is the underlying language data that is used to pre-train the text encoders being in English. Applying the same architectures to another language would require the replacement of the text encoder with one pre-trained explicitly on the target language or trained on multiple languages together.

4.6 Conclusions and Future Work

We have presented two model architectures for detecting sentiment and hate speech in multimodal data (image-text pairs). The comprehensive experimental evaluation of visual, textual and multimodal features for sentiment prediction of (multimodal) tweets has shown that CLIP embeddings can serve as a powerful baseline for multimodal sentiment prediction in tweets. The model architecture evaluated on the MVSA-Single dataset (with tenfold cross-validation) obtained the highest performance among the compared models. A similar pattern is observed by applying a similar methodology for the detection of misogynous memes (that include both text and image content). Our model architecture achieved the best performance (while sharing the results with another model architecture) on the SemEval 2023 Task-5 where the task is to identify the sub-classes of misogyny, such as stereotype, shaming, objectification and violence. We hope this research paves the way for further exploration and development in multimodal sentiment and hate speech detection, ultimately fostering safer and more inclusive digital environments. In future work, we will explore incorporating recent pre-trained large language models and analyse the effect of combining them with visual encoders to perform zero-shot or few-shot experiments.

Acknowledgments The work presented in this paper has received funding from the European Union's Horizon 2020 research and innovation program under the Marie Skłodowska-Curie grant agreement no. 812997 and under the name CLEOPATRA (Cross-lingual Event-centric Open Analytics Research Academy)

References

Alayba AM, Palade V, England M, Iqbal R (2018) A combined CNN and LSTM model for arabic sentiment analysis. In: International Cross-Domain Conference for Machine Learning and Knowledge Extraction, CD-MAKE 2018, Hamburg, August 27–30, 2018, Springer, Lecture

Notes in Computer Science, vol 11015, pp 179–191. https://doi.org/10.1007/978-3-319-99740-7_12

Baziotis C, Pelekis N, Doulkeridis C (2017) Datastories at semeval-2017 task 4: deep LSTM with attention for message-level and topic-based sentiment analysis. In: International Workshop on Semantic Evaluation co-located with Annual Conference of the North American Chapter of the Association for Computational Linguistics: Human Language Technologies, SemEval@NAACL-HLT 2016, San Diego, CA, June 16–17, 2016, The Association for Computer Linguistics, pp 747–754. https://doi.org/10.18653/v1/S17-2126

Borth D, Ji R, Chen T, Breuel TM, Chang S (2013) Large-scale visual sentiment ontology and detectors using adjective noun pairs. In: ACM International Conference on Multimedia, MM 2013, Barcelona, October 21–25, 2013, ACM, pp 223–232. https://doi.org/10.1145/2502081.2502282

Brosch T, Pourtois G, Sander D (2010) The perception and categorisation of emotional stimuli: a review. Cogn Emot 24(3):377–400

Brown TB, Mann B, Ryder N, Subbiah M, Kaplan J, Dhariwal P, Neelakantan A, Shyam P, Sastry G, Askell A, Agarwal S, Herbert-Voss A, Krueger G, Henighan T, Child R, Ramesh A, Ziegler DM, Wu J, Winter C, Hesse C, Chen M, Sigler E, Litwin M, Gray S, Chess B, Clark J, Berner C, McCandlish S, Radford A, Sutskever I, Amodei D (2020) Language models are few-shot learners. In: Annual Conference on Neural Information Processing Systems, NeurIPS 2020, December 6–12, 2020, Virtual Event. https://proceedings.neurips.cc/paper/2020/hash/1457c0d6bfcb4967418bfb8ac142f64a-Abstract.html

Cai G, Xia B (2015) Convolutional neural networks for multimedia sentiment analysis. In: CCF International Conference on Natural Language Processing and Chinese Computing, NLPCC 2015, Nanchang, October 9–13, 2015. Springer, Lecture Notes in Computer Science, vol 9362, pp 159–167. https://doi.org/10.1007/978-3-319-25207-0_14

Cao D, Ji R, Lin D, Li S (2016) A cross-media public sentiment analysis system for microblog. Multimed Syst 22(4):479–486. https://doi.org/10.1007/s00530-014-0407-8

Cheema GS, Hakimov S, Müller-Budack E, Ewerth R (2021) A fair and comprehensive comparison of multimodal tweet sentiment analysis methods. In: Workshop on Multi-Modal Pre-training for Multimedia Understanding co-located with International Conference on Multimedia Retrieval, MMPT@ICMR 2021, August 21, 2021, Virtual Event, ACM, pp 37–45. https://doi.org/10.1145/3463945.3469058

Compton RJ (2003) The interface between emotion and attention: a review of evidence from psychology and neuroscience. Behav Cognit Neurosci Rev 2(2):115–129

Devlin J, Chang M, Lee K, Toutanova K (2019) BERT: pre-training of deep bidirectional transformers for language understanding. In: Conference of the North American Chapter of the Association for Computational Linguistics: Human Language Technologies, NAACL-HLT 2019, Minneapolis, MN, June 2–7, 2019. Association for Computational Linguistics, pp 4171–4186. https://doi.org/10.18653/v1/n19-1423

Du Y, Liu Z, Li J, Zhao WX (2022) A survey of vision-language pre-trained models. In: International Joint Conference on Artificial Intelligence, IJCAI 2022, Vienna, 23–29 July 2022. ijcai.org, pp 5436–5443. https://doi.org/10.24963/ijcai.2022/762

Erhan D, Goodfellow I, Cukierski W, Bengio Y (2013) Challenges in representation learning: facial expression recognition challenge. https://kaggle.com/competitions/challenges-in-representation-learning-facial-expression-recognition-challenge

Fan S, Shen Z, Jiang M, Koenig BL, Xu J, Kankanhalli MS, Zhao Q (2018) Emotional attention: a study of image sentiment and visual attention. In: IEEE Conference on Computer Vision and Pattern Recognition, CVPR 2018, Salt Lake City, UT, June 18–22, 2018. IEEE Computer Society, pp 7521–7531. https://doi.org/10.1109/CVPR.2018.00785. http://openaccess.thecvf.com/content_cvpr_2018/html/Fan_Emotional_Attention_A_CVPR_2018_paper.html

Fersini E, Gasparini F, Rizzi G, Saibene A, Chulvi B, Rosso P, Lees A, Sorensen J (2022) Semeval-2022 task 5: multimedia automatic misogyny identification. In: International Workshop on Semantic Evaluation co-located with Annual Conference of the North American Chapter of the Association for Computational Linguistics, SemEval@NAACL 2022, Seattle, Washington,

July 14–15, 2022. Association for Computational Linguistics, pp 533–549. https://doi.org/10.18653/V1/2022.SEMEVAL-1.74

Gasparini F, Erba I, Fersini E, Corchs S (2018) Multimodal classification of sexist advertisements. In: International Joint Conference on e-Business and Telecommunications, ICETE 2018, Porto, July 26–28, 2018. SciTePress, pp 565–572. https://doi.org/10.5220/0006859405650572

Gomez R, Gibert J, Gómez L, Karatzas D (2020) Exploring hate speech detection in multimodal publications. In: IEEE Winter Conference on Applications of Computer Vision, WACV 2020, Snowmass Village, CO, March 1–5, 2020. IEEE, pp 1459–1467. https://doi.org/10.1109/WACV45572.2020.9093414

Goodfellow IJ, Erhan D, Carrier PL, Courville AC, Mirza M, Hamner B, Cukierski W, Tang Y, Thaler D, Lee D, Zhou Y, Ramaiah C, Feng F, Li R, Wang X, Athanasakis D, Shawe-Taylor J, Milakov M, Park J, Ionescu RT, Popescu M, Grozea C, Bergstra J, Xie J, Romaszko L, Xu B, Chuang Z, Bengio Y (2015) Challenges in representation learning: a report on three machine learning contests. Neur Netw 64:59–63. https://doi.org/10.1016/j.neunet.2014.09.005

Hakimov S, Cheema GS, Ewerth R (2022) TIB-VA at semeval-2022 task 5: a multimodal architecture for the detection and classification of misogynous memes. In: International Workshop on Semantic Evaluation co-located with Annual Conference of the North American Chapter of the Association for Computational Linguistics, SemEval@NAACL 2022, Seattle, Washington, July 14–15, 2022. Association for Computational Linguistics, pp 756–760. https://doi.org/10.18653/v1/2022.semeval-1.105

He K, Zhang X, Ren S, Sun J (2016) Deep residual learning for image recognition. In: IEEE Conference on Computer Vision and Pattern Recognition, CVPR 2016, Las Vegas, NV, June 27–30, 2016. IEEE Computer Society, pp 770–778. https://doi.org/10.1109/CVPR.2016.90

Hossain MZ, Sohel F, Shiratuddin MF, Laga H (2019) A comprehensive survey of deep learning for image captioning. ACM Comput Surv 51(6):118:1–118:36. https://doi.org/10.1145/3295748

Huang M, Cao Y, Dong C (2016) Modeling rich contexts for sentiment classification with LSTM. CoRR abs/1605.01478. http://arxiv.org/abs/1605.01478, 1605.01478

Jiang T, Wang J, Liu Z, Ling Y (2020) Fusion-extraction network for multimodal sentiment analysis. In: Pacific-Asia Conference on Knowledge Discovery and Data Mining, PAKDD 2020, May 11–14, 2020. Virtual Event, Springer, Lecture Notes in Computer Science, vol 12085, pp 785–797. https://doi.org/10.1007/978-3-030-47436-2_59

Karpathy A, Fei-Fei L (2015) Deep visual-semantic alignments for generating image descriptions. In: IEEE Conference on Computer Vision and Pattern Recognition, CVPR 2015, Boston, MA, June 7–12, 2015. IEEE Computer Society, pp 3128–3137. https://doi.org/10.1109/CVPR.2015.7298932

Kiela D, Firooz H, Mohan A, Goswami V, Singh A, Fitzpatrick CA, Bull P, Lipstein G, Nelli T, Zhu R, Muennighoff N, Velioglu R, Rose J, Lippe P, Holla N, Chandra S, Rajamanickam S, Antoniou G, Shutova E, Yannakoudakis H, Sandulescu V, Ozertem U, Pantel P, Specia L, Parikh D (2020a) The hateful memes challenge: Competition report. In: Conference on Neural Information Processing Systems, NeurIPS 2020, 6–12 December 2020, Virtual Event, PMLR, Proceedings of Machine Learning Research, vol 133, pp 344–360. http://proceedings.mlr.press/v133/kiela21a.html

Kiela D, Firooz H, Mohan A, Goswami V, Singh A, Ringshia P, Testuggine D (2020b) The hateful memes challenge: detecting hate speech in multimodal memes. In: Annual Conference on Neural Information Processing Systems NeurIPS 2020, December 6–12, 2020, Virtual Event. https://proceedings.neurips.cc/paper/2020/hash/1b84c4cee2b8b3d823b30e2d604b1878-Abstract.html

Kim Y (2014) Convolutional neural networks for sentence classification. In: Conference on Empirical Methods in Natural Language Processing, EMNLP 2014, Doha, Qatar, October 25–29, 2014. ACL, pp 1746–1751. https://doi.org/10.3115/v1/d14-1181

Kingma DP, Ba J (2015a) Adam: a method for stochastic optimization. In: International Conference on Learning Representations, ICLR 2015, San Diego, CA, May 7–9, 2015. http://arxiv.org/abs/1412.6980

Kingma DP, Ba J (2015b) Adam: a method for stochastic optimization. In: International Conference on Learning Representations, ICLR 2015, San Diego, CA, May 7–9, 2015. http://arxiv.org/abs/1412.6980

LeCun Y, Bengio Y, et al. (1995) Convolutional networks for images, speech, and time series. Handbook Brain Theory Neur. Netw. 3361(10):1995

Li J, Wang Y, Wang C, Tai Y, Qian J, Yang J, Wang C, Li J, Huang F (2019) DSFD: dual shot face detector. In: IEEE Conference on Computer Vision and Pattern Recognition, CVPR 2019, Long Beach, CA, June 16–20, 2019. Computer Vision Foundation/IEEE, pp 5060–5069. https://doi.org/10.1109/CVPR.2019.00520. http://openaccess.thecvf.com/content_CVPR_2019/html/Li_DSFD_Dual_Shot_Face_Detector_CVPR_2019_paper.html

Li X, Yin X, Li C, Zhang P, Hu X, Zhang L, Wang L, Hu H, Dong L, Wei F, Choi Y, Gao J (2020) Oscar: object-semantics aligned pre-training for vision-language tasks. In: European Conference on Computer Vision, August 23–28, 2020. Virtual Event, Springer, Lecture Notes in Computer Science, vol 12375, pp 121–137. https://doi.org/10.1007/978-3-030-58577-8_8

Li Z, Xu B, Zhu C, Zhao T (2022) CLMLF: a contrastive learning and multi-layer fusion method for multimodal sentiment detection. In: Annual Conference of the North American Chapter of the Association for Computational Linguistics, NAACL 2022, Seattle, WA, July 10–15, 2022. Association for Computational Linguistics, pp 2282–2294. https://doi.org/10.18653/v1/2022.findings-naacl.175

Liu B, Zhang L (2012) A survey of opinion mining and sentiment analysis. In: Mining Text Data, Springer, pp 415–463. https://doi.org/10.1007/978-1-4614-3223-4_13

Liu Y, Ott M, Goyal N, Du J, Joshi M, Chen D, Levy O, Lewis M, Zettlemoyer L, Stoyanov V (2019) Roberta: a robustly optimized BERT pretraining approach. CoRR abs/1907.11692. http://arxiv.org/abs/1907.11692, 1907.11692

Lu J, Batra D, Parikh D, Lee S (2019) Vilbert: Pretraining task-agnostic visiolinguistic representations for vision-and-language tasks. In: Annual Conference on Neural Information Processing Systems, NeurIPS 2019, Vancouver, BC, December 8–14, 2019, pp 13–23. https://proceedings.neurips.cc/paper/2019/hash/c74d97b01eae257e44aa9d5bade97baf-Abstract.html

Mansimov E, Parisotto E, Ba LJ, Salakhutdinov R (2016) Generating images from captions with attention. In: International Conference on Learning Representations, ICLR 2016, San Juan, Puerto Rico, May 2–4, 2016. http://arxiv.org/abs/1511.02793

Medhat W, Hassan A, Korashy H (2014) Sentiment analysis algorithms and applications: a survey. Ain Shams Eng J 5(4):1093–1113

Menini S, Aprosio AP, Tonelli S (2020) A multimodal dataset of images and text to study abusive language. In: Italian Conference on Computational Linguistics, CLIC-IT 2020, March 1–3, 2021, Virtual Event, CEUR-WS.org, CEUR Workshop Proceedings, vol 2769. http://ceur-ws.org/Vol-2769/paper_11.pdf

Niu T, Zhu S, Pang L, El-Saddik A (2016) Sentiment analysis on multi-view social data. In: International Conference on Multimedia Modeling, MMM 2016, Miami, FL, January 4–6, 2016, Springer, Lecture Notes in Computer Science, vol 9517, pp 15–27. https://doi.org/10.1007/978-3-319-27674-8_2

Pennington J, Socher R, Manning CD (2014) Glove: global vectors for word representation. In: Conference on Empirical Methods in Natural Language Processing, EMNLP 2014, Doha, Qatar, October 25–29, 2014, ACL, pp 1532–1543. https://doi.org/10.3115/v1/d14-1162

Plan International (2020) Free to be online? https://plan-international.org/publications/free-to-be-online/

Poria S, Cambria E, Hazarika D, Majumder N, Zadeh A, Morency L (2017) Context-dependent sentiment analysis in user-generated videos. In: Annual Meeting of the Association for Computational Linguistics, ACL 2017, Vancouver, July 30–August 4, 2017, Association for Computational Linguistics, pp 873–883. https://doi.org/10.18653/v1/P17-1081

Pramanick S, Dimitrov D, Mukherjee R, Sharma S, Akhtar MS, Nakov P, Chakraborty T (2021) Detecting harmful memes and their targets. In: Annual Meeting of the Association for Computational Linguistics and International Joint Conference on Natural Language Processing,

ACL-IJCNLP 2021, August 1–6, 2021, Virtual Event, Association for Computational Linguistics, Findings of ACL, vol ACL-IJCNLP 2021, pp 2783–2796. https://doi.org/10.18653/v1/2021.findings-acl.246

Qiao T, Zhang J, Xu D, Tao D (2019) Mirrorgan: learning text-to-image generation by redescription. In: IEEE Conference on Computer Vision and Pattern Recognition, CVPR 2019, Long Beach, CA, June 16–20, 2019, Computer Vision Foundation/IEEE, pp 1505–1514. https://doi.org/10.1109/CVPR.2019.00160

Radford A, Kim JW, Hallacy C, Ramesh A, Goh G, Agarwal S, Sastry G, Askell A, Mishkin P, Clark J, Krueger G, Sutskever I (2021) Learning transferable visual models from natural language supervision. In: International Conference on Machine Learning, ICML 2021, 18–24 July 2021, Virtual Event, PMLR, Proceedings of Machine Learning Research, vol 139, pp 8748–8763. http://proceedings.mlr.press/v139/radford21a.html

Ramesh A, Pavlov M, Goh G, Gray S, Voss C, Radford A, Chen M, Sutskever I (2021) Zero-shot text-to-image generation. In: International Conference on Machine Learning, ICML 2021, 18–24 July 2021, Virtual Event, PMLR, Proceedings of Machine Learning Research, vol 139, pp 8821–8831. http://proceedings.mlr.press/v139/ramesh21a.html

Rumelhart DE, Hinton GE, Williams RJ (1986) Learning internal representations by error propagation. In: Parallel Distributed Processing: Explorations in the Microstructure of Cognition, vol 1: Foundations, pp 318–362

Russakovsky O, Deng J, Su H, Krause J, Satheesh S, Ma S, Huang Z, Karpathy A, Khosla A, Bernstein MS, Berg AC, Li F (2015) Imagenet large scale visual recognition challenge. Int J Comput Vision 115(3):211–252. https://doi.org/10.1007/s11263-015-0816-y

Sharma C, Bhageria D, Scott W, PYKL S, Das A, Chakraborty T, Pulabaigari V, Gambäck B (2020) Semeval-2020 task 8: memotion analysis- the visuo-lingual metaphor! In: Workshop on Semantic Evaluation co-located with International Conference on Computational Linguistics, SemEval@COLING 2020, December 12–13, 2020, Virtual Event, International Committee for Computational Linguistics, pp 759–773. https://doi.org/10.18653/v1/2020.semeval-1.99

Shin B, Lee T, Choi JD (2017) Lexicon integrated CNN models with attention for sentiment analysis. In: Workshop on Computational Approaches to Subjectivity, Sentiment and Social Media Analysis co-located with Conference on Empirical Methods in Natural Language Processing, WASSA@EMNLP 2017, Copenhagen, Denmark, September 8, 2017, Association for Computational Linguistics, pp 149–158. https://doi.org/10.18653/v1/w17-5220

Simonyan K, Zisserman A (2015) Very deep convolutional networks for large-scale image recognition. In: International Conference on Learning Representations, ICLR 2015, San Diego, CA, May 7–9, 2015. http://arxiv.org/abs/1409.1556

Smeulders AWM, Worring M, Santini S, Gupta A, Jain RC (2000) Content-based image retrieval at the end of the early years. IEEE Trans Pattern Analy Mach Intell 22(12):1349–1380. https://doi.org/10.1109/34.895972

Socher R, Karpathy A, Le QV, Manning CD, Ng AY (2014) Grounded compositional semantics for finding and describing images with sentences. Trans Assoc Comput Linguist 2:207–218. https://tacl2013.cs.columbia.edu/ojs/index.php/tacl/article/view/325

Soleymani M, García D, Jou B, Schuller BW, Chang S, Pantic M (2017) A survey of multimodal sentiment analysis. Image Vision Comput 65:3–14. https://doi.org/10.1016/j.imavis.2017.08.003

Suryawanshi S, Chakravarthi BR, Arcan M, Buitelaar P (2020) Multimodal meme dataset (multioff) for identifying offensive content in image and text. In: Workshop on Trolling, Aggression and Cyberbullying Co-located with International Conference on Language Resources and Evaluation, TRAC@LREC 2020, May 2020, Virtual Event, European Language Resources Association (ELRA), pp 32–41. https://aclanthology.org/2020.trac-1.6/

Thelwall M, Buckley K, Paltoglou G, Cai D, Kappas A (2010) Sentiment in short strength detection informal text. J Amer Soc Inf Sci Technol 61(12):2544–2558. https://doi.org/10.1002/asi.21416

Vaswani A, Shazeer N, Parmar N, Uszkoreit J, Jones L, Gomez AN, Kaiser L, Polosukhin I (2017) Attention is all you need. In: Annual Conference on Neural Information Processing Systems,

NIPS 2017, Long Beach, CA, December 4–9, 2017, pp 5998–6008. https://proceedings.neurips.cc/paper/2017/hash/3f5ee243547dee91fbd053c1c4a845aa-Abstract.html

Vogels EA (2021) The State of Online Harassment. Pew Research Center. https://www.pewresearch.org/internet/2021/01/13/the-state-of-online-harassment/

Xu N, Mao W (2017) Multisentinet: a deep semantic network for multimodal sentiment analysis. In: ACM on Conference on Information and Knowledge Management, CIKM 2017, Singapore, November 06–10, 2017, ACM, pp 2399–2402. https://doi.org/10.1145/3132847.3133142

Xu N, Mao W, Chen G (2018) A co-memory network for multimodal sentiment analysis. In: International ACM Conference on Research & Development in Information Retrieval, SIGIR 2018, Ann Arbor, MI, July 08–12, 2018, ACM, pp 929–932. https://doi.org/10.1145/3209978.3210093

You Q, Luo J, Jin H, Yang J (2016) Building a large scale dataset for image emotion recognition: the fine print and the benchmark. In: AAAI Conference on Artificial Intelligence, AAAI 2016, Phoenix, Arizona, February 12–17, 2016, AAAI Press, pp 308–314. http://www.aaai.org/ocs/index.php/AAAI/AAAI16/paper/view/12272

Yu Y, Lin H, Meng J, Zhao Z (2016) Visual and textual sentiment analysis of a microblog using deep convolutional neural networks. Algorithms 9(2):41. https://doi.org/10.3390/a9020041

Zeinert P, Inie N, Derczynski L (2021) Annotating online misogyny. In: Annual Meeting of the Association for Computational Linguistics and International Joint Conference on Natural Language Processing, ACL-IJCNLP 2021, August 1–6, 2021, Virtual Event, Association for Computational Linguistics, pp 3181–3197. https://doi.org/10.18653/v1/2021.acl-long.247

Zhen L, Hu P, Wang X, Peng D (2019) Deep supervised cross-modal retrieval. In: IEEE Conference on Computer Vision and Pattern Recognition, CVPR 2019, Long Beach, CA, June 16–20, 2019, Computer Vision Foundation/IEEE, pp 10394–10403. https://doi.org/10.1109/CVPR.2019.01064. http://openaccess.thecvf.com/content_CVPR_2019/html/Zhen_Deep_Supervised_Cross-Modal_Retrieval_CVPR_2019_paper.html

Zhou B, Lapedriza À, Khosla A, Oliva A, Torralba A (2018) Places: a 10 million image database for scene recognition. IEEE Trans Pattern Analy Mach Intell 40(6):1452–1464. https://doi.org/10.1109/TPAMI.2017.2723009

Chapter 5
Effect of Unknown and Fragmented Tokens on the Performance of Multilingual Language Models at Low-Resource Tasks

Arijit Nag, Bidisha Samanta, Animesh Mukherjee, Niloy Ganguly, and Soumen Chakrabarty

Abstract Multilingual language models (MLLMs) like mBERT promise to extend the benefits of NLP research to low-resource languages (LRLs). However, LRL vocabulary is often seriously under-represented in the workpiece dictionaries of MLLMs. This leads to many LRL words being replaced by UNK (unknown tokens) or concatenated from morphologically unrelated wordpieces, consequently leading to low task accuracy. Pre-training MLLMs after including LRL documents is extremely resource-intensive in terms of both human inputs and computational resources. In this chapter, we study intuitive strategies to seek and protect "vulnerable words" in LRLs by introducing them into MLLM dictionaries, providing reasonable initialisations of their embeddings, followed by limited fine-tuning, subject to the limits of available LRL corpora. Our experiments show some significant performance improvements and also some surprising limits to such vocabulary augmentation strategies in various classification tasks in multiple diverse LRLs as well as code-mixed datasets. We release the code and data to enable further research.

A. Nag (✉) · B. Samanta · A. Mukherjee · N. Ganguly
IIT Kharagpur, Kharagpur, India
e-mail: arijitnag@iitkgp.ac.in; bidisha@iitkgp.ac.in; animeshm@iitkgp.ac.in; niloy@cse.iitkgp.ac.in

S. Chakrabarti
IIT Bombay, Bombay, India
e-mail: soumen@cse.iitb.ac.in

I. Marenzi et al. (eds.), *Event Analytics across Languages and Communities*,
https://doi.org/10.1007/978-3-031-64451-1_5

5.1 Introduction

It is a common practice to start with a multilingual language model (MLLM) like mBERT,[1] XLM-R (Conneau et al. 2020), etc., which has been pre-trained with large multilingual corpora, and fine-tune the MLLM for diverse downstream tasks. MLLMs support many low-resource languages (LRLs) via universal character sets and wordpiece vocabularies. However, closer inspection of these MLLMs reveals that the portion of vocabulary allotted to LRLs can be orders of magnitude lower than that allotted to high-resource languages (HRLs) such as English, as shown in Table 5.1. Partly due to this imbalance, MLLM performance for different downstream tasks for LRLs lags behind HRLs. The problem is more severe for an LRL that is typologically dissimilar from any HRL represented well in the MLLM's vocabulary. Words from such language groups will likely get over-segmented or, in the worst case, can be conflated to the UNK token. One way to solve this issue is to pre-train a LM with a large corpus in the target language. By definition, collecting such large corpora for LRLs is both difficult and expensive. Furthermore, this approach also requires expensive computation.

In this work, we study the effectiveness of vocabulary augmentation on an existing MLLM during the fine-tuning stage for various downstream classification tasks covering multiple LRLs and also a code-mixed language. We gradually increase the number of newly added words to the existing MLLM dictionary, starting with the words that map to UNK or are most severely fragmented after tokenisation. We define a simple surrogate for the amount of fragmentation and include words with decreasing estimated fragmentation. Our study shows that, for each of the datasets covering multiple LRLs, augmenting with words that map to UNK or over-fragmented words helps improve the performance rather than augmenting word fragments, which has recently been explored (Hong et al. 2021). Surprisingly, continuous addition of words with decreasing fragmentation does not markedly improve model performance. For the initialisation of embeddings of the newly added LRL words, we experiment with three different strategies, including

Table 5.1 Representation of the vocabulary of various Indian languages in mBERT's wordpiece dictionary. *Based on basic to extended Latin script Unicode range

Language	Vocab count	Percentage (%)
Bengali	946	0.79
Hindi	1,852	1.55
Gujarati	404	0.34
Kannada	653	0.55
Malayalam	565	0.47
Tamil	832	0.7
Telugu	887	0.74
English*	64,529–78,984	53.98–66.07

[1] https://github.com/google-research/bert/blob/master/multilingual.md.

Algorithm 1 LRL vocabulary selection

inputs LRL corpus D, MLLM tokeniser $\mathcal{T}$, token fragmentation threshold $\theta < 1$
outputs V_{UNK}: LRL words to include that were mapped to UNK, V_{frag}: LRL words to include that were excessively fragmented

$V_{\text{UNK}} \leftarrow$ empty list
$V_{\text{frag}} \leftarrow$ empty list
$\mathcal{W} \leftarrow$ distinct words from corpus D
for $w \in \mathcal{W}$ **do**
    if $\mathcal{T}(w) =$ UNK **then**
        add w to V_{UNK}
    else
        tokens $\leftarrow \mathcal{T}(w)$
        fragment $\leftarrow$ |tokens| / chars(w)
        if fragment $\geq \theta$ **then**
            add w to V_{frag}
return V_{UNK}, V_{frag}

taking clues from the same language to transfer embedding from a high-resource target language. The result suggests a mixture of a particular LRL and an HRL (in our case, English) works the best.

5.2 Related Work

Continued pre-training (Tai et al. 2020; Ebrahimi and Kann 2021; Wang et al. 2020; Chau et al. 2020) with or without adding extra vocabulary to existing LMs like monolingual BERT, multilingual BERT (mBERT), XLM-R, etc. proves beneficial for improving domain and language-specific performances over various tasks. Some works (Ruzzetti et al. 2021; Yu et al. 2021) try to incorporate dictionaries to enhance the performance of LMs by focusing on rare and OOV (Out of Vocabulary) words. Liu et al. (2021) propose an embedding generator module in the pretrain-fine-tune pipeline to resolve the vocabulary gaps in the pretrain-fine-tune paradigm. Adaptors (Sachidananda et al. 2021; Moon and Okazaki 2020; Hofmann et al. 2021) are also showing promising outcomes in LRL modelling. Chung et al. (2020) explore multilingual vocabulary generation from language clusters. Minixhofer et al. (2021) propose a technique to transfer English language models to new languages without expensive computation. Others (Wang et al. 2019; Hong et al. 2021) focus on embedding initialisation for newly added vocabulary words, which are word fragments, which is also among our concerns. We compare our performance in detail with AVocaDo (Hong et al. 2021) in the experiment section.

5.3 Methods

Our method has two key steps. In the first step (Sect. 5.3.1), we select a subset of candidate LRL words to add from a downstream task dataset (typically, only the train fold). In the second step (Sect. 5.3.2), we initialise the embeddings of the newly introduced LRL words. In Sect. 5.3.3, we provide some further details about regularisation of embeddings (Table 5.2).

5.3.1 Vulnerable LRL Word Selection

This section describes our methodology for selecting the subset of words to add to the existing model through Algorithm 1. Our intention here is to choose the words that cannot be assembled from wordpieces by the MLLM tokeniser or can be assembled from an unacceptably large number of fragments. Such over-fragmentation may adversely affect downstream task performance, if the fragmentation is not morphologically meaningful or if there is not enough data for the MLLM to learn how to assemble wordpiece embeddings into LRL word embeddings.

In general, detecting when an LRL word fragmentation is benign and when it is dangerous is extremely difficult. We use a simple surrogate that is intuitive and can be computed very fast. We tokenise the LRL word with the existing MLLM tokeniser and find the length of the output token list. Our surrogate measure is the ratio of the length of the token list to the number of characters in the original LRL word. If the ratio is high, then the existing tokeniser is likely over-segmenting the word. We also want to include words that the MLLM tokeniser cannot assemble

Table 5.2 Salient statistics of tasks. Note the small size of LRL datasets

	Tasks	Language	Train instances	Test instances
(a)	IITP Product Review (Kakwani et al. 2020)	Hindi	4,182	523
(b)	Bengali Sentiment Analysis (Islam et al. 2021)	Bengali	12,576	1,587
(c)	Bengali HateSpeech (Karim et al. 2020)	Bengali	981	295
(d)	Gujarati headline classification (Arora 2020)	Gujarati	5,269	659
(e)	Malayalam headline classification (Arora 2020)	Malayalam	5,036	630
(f)	GLUECoS Sentiment Analysis (Khanuja et al. 2020)	Hindi-English code-mix	10,079	1,260

from wordpieces and marks as UNK. Such words may be even more vulnerable than over-segmented words, because all such words get conflated to the same non-contextual ('UNK') embedding.

5.3.2 *Embedding Initialisation*

Here we describe the different ways to initialise the embeddings of newly added words.

InitLRL: The embedding of the newly introduced LRL word is initialised using other LRL wordpieces already in the MLLM dictionary. Suppose we add the Bengali word 'হাসপাতাল', ('hospital' in English). Suppose the existing MLLM tokeniser splits it into ['হ', '##◌াস', '##প', '##◌াত', '##◌াল']. Then we initialise the embedding of ' 'হাসপাতাল' ' with the average of the existing MLLM embeddings of the fragments in this list.

InitHRL: We translate the newly introduced LRL word into an HRL (English in our experiments) and initialise the embedding of the LRL word using embeddings of existing HRL tokens in the MLLM dictionary. Continuing the above example, we initialise the embedding of ' 'হাসপাতাল'' with the average of the existing MLLM embeddings of the wordpieces into which "hospital" is decomposed. Note that InitHRL is the only option available to words in V_{UNK}.

InitMix: We use the average of InitLRL and InitHRL embeddings.

5.3.3 *Fine-Tuning and Regularisation*

It is challenging to learn good contextual embedding for words in $V_{\text{UNK}} \cup V_{\text{frag}}$ due to very small task-specific training data compared to the MLLM pre-training corpus. Therefore, we found it necessary to apply some regularisation to avoid overfitting during fine-tuning. Let $\mathcal{T}, \mathcal{T}'$ be the initial and final MLLM tokenisers. For a particular sentence $S = w_1, w_2, \ldots, w_I$ with words w_i, we will get two different tokenisations; these will generally lead to different contextual embeddings $E = (e_1, \ldots, e_K)$ and $E' = (e'_1, \ldots, e'_L)$; generally $K \neq L$. We average-pool these to get vectors $\boldsymbol{e}, \boldsymbol{e}'$, which a final layer uses for the classification task, with losses $\ell_{\mathcal{T}}$ and $\ell_{\mathcal{T}'}$. We also use $(\boldsymbol{e} + \boldsymbol{e}')/2$ for a third classification, with loss ℓ_{mix}. The overall training loss is $\ell_{\mathcal{T}} + \ell_{\mathcal{T}'} + \ell_{\text{mix}}$, where $\ell_{\mathcal{T}}$ and ℓ_{mix} are expected to reduce overfitting (Table 5.3).

Table 5.3 Added vocab count with different setups for all tasks. Here UNK $= |V_{\text{UNK}}|$, and UXXX denotes $|V_{\text{UNK}} \cup V_{\text{frag}}|$ with $\theta = \text{XXX}/100$

	Added vocab size						
Task	UNK	U100	U090	U080	U070	U060	U050
(a)	156	598	599	1,311	2,454	5,027	6,863
(b)	2,806	4,574	4,582	7,053	10,433	16,712	21,118
(c)	361	686	687	963	1,455	2,339	2,917
(d)	2	1,462	1,485	3,817	6,172	8,732	9,921
(e)	2	475	513	1,547	3,513	6,602	9,805
(f)	28	960	964	1,849	3,480	6,359	8,146

5.4 Experiments

In this section, we discuss the experimental results, the experiment settings, the dataset used for experiments and the evaluation metrics.

5.4.1 Experimental Settings

In all experiments, we trained the models on a single NVIDIA TITAN X with 12GB of memory. We implemented all models with PyTorch using the Transformers library from Hugging Face. Our model has ∼29M trainable parameters, and it takes 20–45 minutes to train depending on the size of datasets. Details of model hyperparameters are present in Table 5.4.

5.4.2 Datasets and Evaluation Metric

We experiment with six short multi-class text classification tasks covering four Indian languages and a Hindi-English code-mixed dataset. We show the details of the datasets in Table 5.2. We use mBERT as the MLLM. $|V_{\text{UNK}}|$ and $|V_{\text{UNK}} \cup V_{\text{frag}}|$ for various thresholds $\theta \in \{1, .9, .8, .7, .6, .5\}$ are shown in Table 5.3. We report the macro-F1 (Fig. 5.1) and accuracy (Fig. 5.2) score for each task. All the results are averaged over three random seeds.

Table 5.4 Hyperparameters used in experiments. We find the best hyperparameter settings using manual search according to macro-f1 performance

Hyperparameter	Value
mBERT version	Bert-base-multilingual-cased
Batch size	16.32
Epoch	10
Learning rate	1×10^{-5}
max_seq_len	128

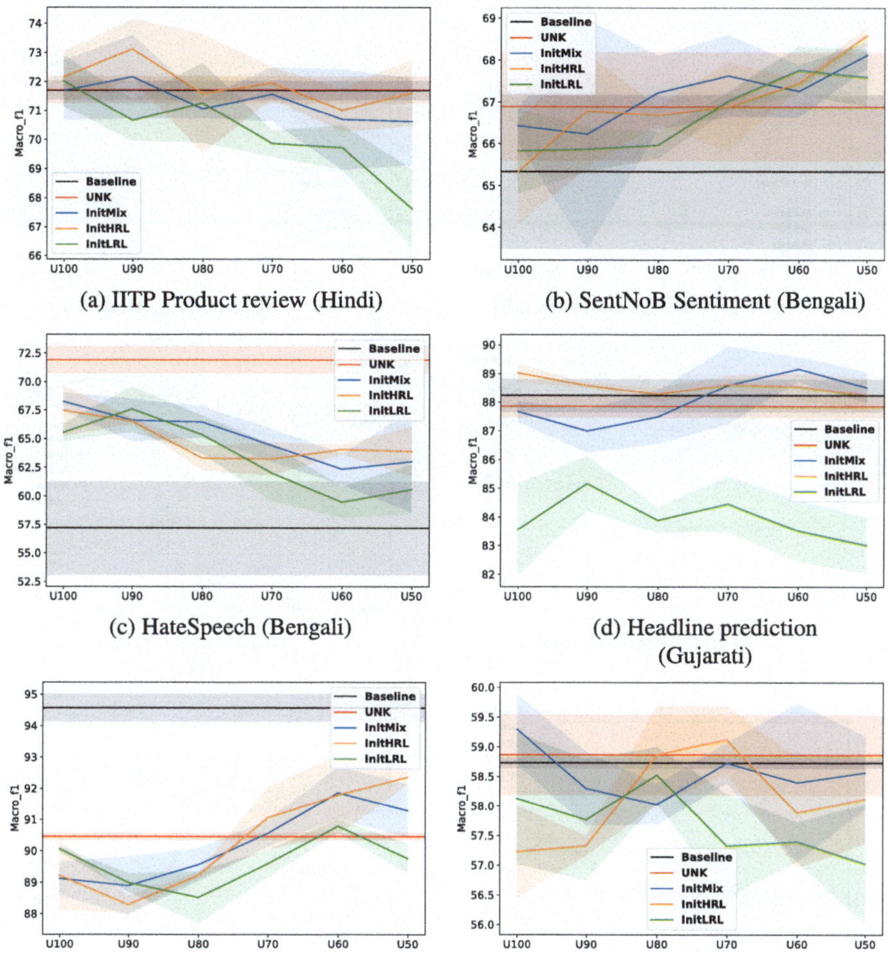

(a) IITP Product review (Hindi)

(b) SentNoB Sentiment (Bengali)

(c) HateSpeech (Bengali)

(d) Headline prediction (Gujarati)

(e) Headline prediction (Malayalam)

(f) GLUECoS (Hindi-English (code-mix)

Fig. 5.1 Macro F1 vs. increasing LRL words added to MLLM dictionary. Black = MLLM baseline. Red = augment with only V_{UNK}. The blue, orange and green line represents the performance with V_{UNK} and V_{frag} with different levels of θ and embedding initialisation. Solid lines = average. Shaded region shows the standard deviation over five random runs

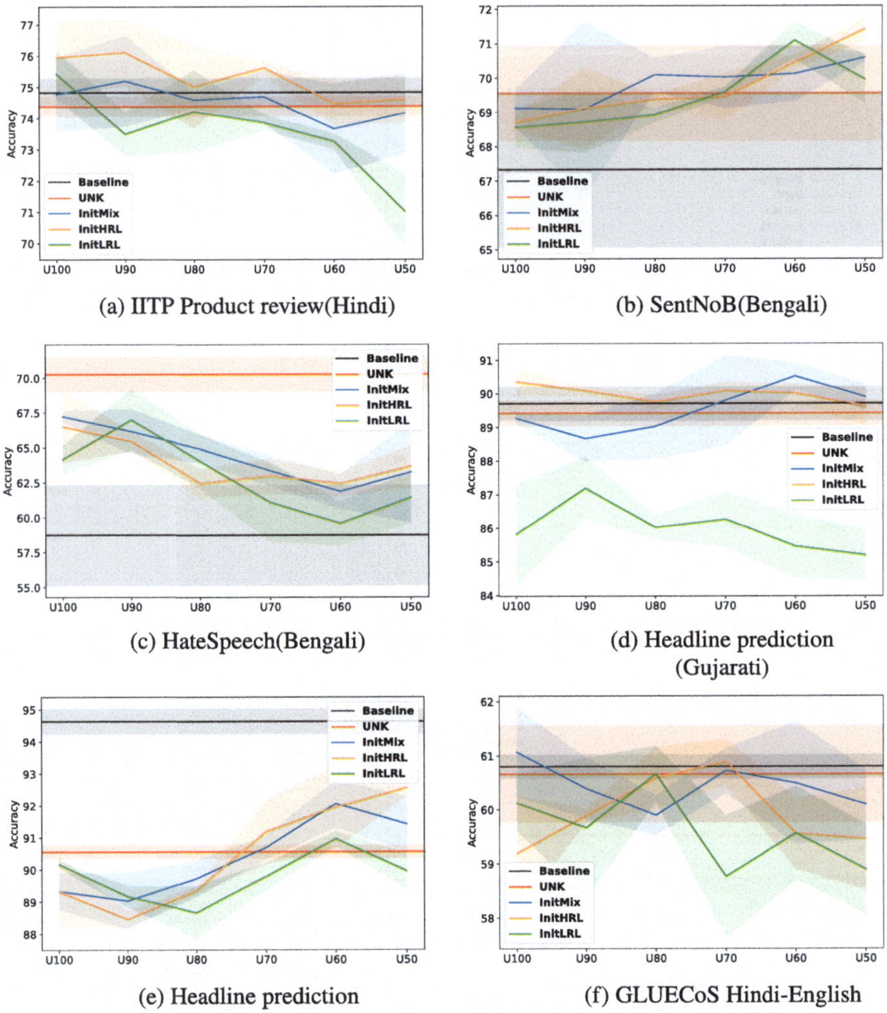

(a) IITP Product review(Hindi)

(b) SentNoB(Bengali)

(c) HateSpeech(Bengali)

(d) Headline prediction (Gujarati)

(e) Headline prediction (Malayalam)

(f) GLUECoS Hindi-English (code-mix)

Fig. 5.2 Accuracy vs. increasing LRL words added to MLLM dictionary. Black = MLLM baseline. Red = augment with only V_{UNK}. The blue, orange and green line represents the performance with V_{UNK} and V_{frag} with different levels of θ and embedding initialisation. Solid lines = average. Shaded region shows the standard deviation over five random runs

5.4.3 *Quantitative Results*

In Fig. 5.1, we plot task macro-F1 for different extents of vocabulary augmentation (as in Table 5.3) for all tasks. Green, orange and blue lines show the performance after adding $V_{\text{UNK}} \cup V_{\text{frag}}$ for various θ with InitLRL, InitHRL and InitMix initialisation, respectively. Similarly coloured bands show 1-standard deviation spreads.

$V_{\text{UNK}} \cup V_{\text{frag}}$ **helps:** For all tasks except Malayalam headline classification, performance after adding $V_{\text{UNK}} \cup V_{\text{frag}}$ is always better than baseline MLLM, and the gap is usually significant. This shows that even sparse training of LRL tokens that used to be UNK or over-fragmented to the MLLM is helping to improve model performance.

More Augmentation≠Performance Boost We expected that additional non-UNK vulnerable words in V_{frag} might further improve task performance, but surprisingly, this was not universally the case. A growing V_{frag} does improve the performance for SentNoB and Malayalam headline prediction datasets, but the gain is not always consistent or significant.

Reliable Initialisation Matters If we initialise the embeddings of new LRL words using InitHRL or InitMix, performance is better than using InitLRL. Transfer of embeddings from a well-represented HRL is the likely reason.

Comparison with AVocaDo In Table 5.5, results show that AVocaDo-style vocabulary augmentation causes performance degradation for all the LRL datasets, and the performance gap between the best-performing model and AVocaDo slims out when no additional vocabulary is augmented. Along with these, for most of the datasets, there are performance improvements of AVocaDo{U_100} over AVocaDo{X} due to the fact that it generates more fragmented words (details in Table 5.6) as tokens; and as LRLs are under-represented in the existing MLMs, initialisation for the new tokens is challenging.

Table 5.5 Here AVocaDo_0, AVocaDo{U_100} and AVocaDo{X} represent the AVocaDo's performance with zero, U_100 and AVocaDo style ~|U_100| of vocab augmentation, respectively. Ours{U_100} shows our best-performing model's performance with U_100. (a)–(f) are the datasets/tasks defined in Table 5.2

Tasks→	(a)	(b)	(c)	(d)	(e)	(f)
AVocaDo_0	70.56	65.97	64.52	88.00	93.92	58.41
Ours{U_100}	72.17	66.43	68.27	89.03	90.07	59.30
AVocaDo{U_100}	69.87	62.53	57.67	87.56	92.83	56.16
AVocaDo{X}	66.35	64.98	47.40	85.23	93.89	55.59

Table 5.6 Here we compare the average token length of the tokens in {U_100} and AVocaDo-style ∼|U_100| of tokens. It shows except one for all the cases AVocaDo generate smaller tokens than U_100. (a)–(f) are the datasets/tasks defined in Table 5.2

Tasks→	(a)	(b)	(c)	(d)	(e)	(f)
Ours{U_100}	3.69	5.41	4.72	4.09	4.16	3.17
AVocaDo{X}	3.42	4.18	3.38	3.62	3.56	3.49

Table 5.7 Ablation: The first and second rows show our best model performance, trained with all three losses and only $\ell_{\mathcal{T}'}$, respectively. The last row is for baseline mBERT without dictionary augmentation

Tasks→	(a)	(b)	(c)	(d)	(e)	(f)
Our method	73.13	68.60	71.93	89.17	92.37	59.13
$-\{\ell_{\mathcal{T}}, \ell_{\text{mix}}\}$	70.50	67.67	66.37	89.03	91.87	57.30
mBERT	71.70	65.33	57.17	88.23	94.57	58.73

5.4.4 Ablation

In Table 5.7, we show our work's effectiveness with regularisation components. Our model with ℓ_{mix}, $\ell_{\mathcal{T}}$ and $\ell_{\mathcal{T}'}$ loss components beats the base mBERT model for all, except the Malayalam headline dataset. There is usually a drop in performance if we discard the ℓ_{mix} and $\ell_{\mathcal{T}}$ loss components.

5.5 Conclusion and Future Work

In this chapter, we study the effects of under-represented words in mBERT for multiple low-resource Indian language classification datasets. Our study reveals the positive impact of adding V_{UNK} to mBERT's vocabulary before fine-tuning. However, further augmentation with V_{frag} had limited or no impact. It is possible that a more sophisticated LRL word selection method may obtain further gains from V_{frag}; this requires further study. We show that reliable initialisation of newly introduced LRL words is important. We also show that loss regularisation is crucial to prevent overfitting new LRL embeddings during fine-tuning, because the LRL task corpus is generally very small compared to the pre-training corpus. In the future, we want to extend the study to other target tasks (especially language generation) and LRLs.

5.6 Limitations

Our work opens up several avenues to explore. Ideally, we want a recipe to inspect an LRL task and corpus, collect some salient statistics and roughly predict the benefits of MLLM dictionary augmentation. Our tasks showed a range of behaviours in this regard, but our understanding of that spectrum is far from complete. Further experiments with mBERT and other MLLMs are warranted. Fragmentation is a simplistic view and a crude approximation of the potential damage that the MLLM might do by attempting to assemble embeddings of vulnerable LRL words from wordpieces. A more sophisticated estimate of vulnerability may show more consistent benefits from V_{frag}.

References

Arora G (2020) iNLTK: Natural language toolkit for Indic languages. In: Proceedings of Second Workshop for NLP Open Source Software (NLP-OSS). Association for Computational Linguistics, pp 66–71, Online. https://doi.org/10.18653/v1/2020.nlposs-1.10, https://www.aclweb.org/anthology/2020.nlposs-1.10

Chau EC, Lin LH, Smith NA (2020) Parsing with multilingual BERT, a small corpus, and a small treebank. In: Findings of the Association for Computational Linguistics: EMNLP 2020. Association for Computational Linguistics, pp 1324–1334, Online. https://doi.org/10.18653/v1/2020.findings-emnlp.118, https://aclanthology.org/2020.findings-emnlp.118

Chung HW, Garrette D, Tan KC, Riesa J (2020) Improving multilingual models with language-clustered vocabularies. In: Proceedings of the 2020 Conference on Empirical Methods in Natural Language Processing (EMNLP). Association for Computational Linguistics, pp 4536–4546, Online. https://doi.org/10.18653/v1/2020.emnlp-main.367, https://aclanthology.org/2020.emnlp-main.367

Conneau A, Khandelwal K, Goyal N, Chaudhary V, Wenzek G, Guzmán F, Grave E, Ott M, Zettlemoyer L, Stoyanov V (2020) Unsupervised cross-lingual representation learning at scale. In: Proceedings of the 58th Annual Meeting of the Association for Computational Linguistics. Association for Computational Linguistics, pp 8440–8451, Online. https://doi.org/10.18653/v1/2020.acl-main.747, https://aclanthology.org/2020.acl-main.747

Ebrahimi A, Kann K (2021) How to adapt your pretrained multilingual model to 1600 languages. In: Proceedings of the 59th Annual Meeting of the Association for Computational Linguistics and the 11th International Joint Conference on Natural Language Processing (Volume 1: Long Papers). Association for Computational Linguistics, pp 4555–4567, Online. https://doi.org/10.18653/v1/2021.acl-long.351, https://aclanthology.org/2021.acl-long.351

Hofmann V, Pierrehumbert J, Schütze H (2021) Superbizarre is not superb: Derivational morphology improves BERT's interpretation of complex words. In: Proceedings of the 59th Annual Meeting of the Association for Computational Linguistics and the 11th International Joint Conference on Natural Language Processing (Volume 1: Long Papers). Association for Computational Linguistics, pp 3594–3608, Online. https://doi.org/10.18653/v1/2021.acl-long.279, https://aclanthology.org/2021.acl-long.279

Hong J, Kim T, Lim H, Choo J (2021) AVocaDo: Strategy for adapting vocabulary to downstream domain. In: Proceedings of the 2021 Conference on Empirical Methods in Natural Language Processing. Association for Computational Linguistics, Online and Punta Cana, Dominican Republic, pp 4692–4700. https://doi.org/10.18653/v1/2021.emnlp-main.385, https://aclanthology.org/2021.emnlp-main.385

Islam KI, Kar S, Islam MS, Amin MR (2021) SentNoB: A dataset for analysing sentiment on noisy Bangla texts. In: Findings of the Association for Computational Linguistics: EMNLP 2021. Association for Computational Linguistics, pp 3265–3271. https://doi.org/10.18653/v1/2021.findings-emnlp.278, https://aclanthology.org/2021.findings-emnlp.278

Kakwani D, Kunchukuttan A, Golla S, NC G, Bhattacharyya A, Khapra MM, Kumar P (2020) IndicNLPSuite: Monolingual Corpora, Evaluation Benchmarks and Pre-trained Multilingual Language Models for Indian Languages. In: Findings of the Association for Computational Linguistics: EMNLP 2020, Association for Computational Linguistics, pp 4948–4961, Online. https://doi.org/10.18653/v1/2020.findings-emnlp.445, https://aclanthology.org/2020.findings-emnlp.445

Karim MR, Chakravarti BR, McCrae JP, Cochez M (2020) Classification benchmarks for under-resourced Bengali language based on multichannel convolutional-LSTM network. In: 7th IEEE International Conference on Data Science and Advanced Analytics (IEEE DSAA,2020). https://doi.org/10.48550/arXiv.2004.07807, https://arxiv.org/abs/2004.07807

Khanuja S, Dandapat S, Srinivasan A, Sitaram S, Choudhury M (2020) GLUECoS: An evaluation benchmark for code-switched NLP. In: Proceedings of the 58th Annual Meeting of the Association for Computational Linguistics. Association for Computational Linguistics, pp 3575–3585, Online. https://www.aclweb.org/anthology/2020.acl-main.329

Liu X, Yang B, Liu D, Zhang H, Luo W, Zhang M, Zhang H, Su J (2021) Bridging subword gaps in pretrain-finetune paradigm for natural language generation. In: Proceedings of the 59th Annual Meeting of the Association for Computational Linguistics and the 11th International Joint Conference on Natural Language Processing (Volume 1: Long Papers). Association for Computational Linguistics, pp 6001–6011, Online. https://doi.org/10.18653/v1/2021.acl-long.468, https://aclanthology.org/2021.acl-long.468

Minixhofer B, Paischer F, Rekabsaz N (2021) Wechsel: Effective initialization of subword embeddings for cross-lingual transfer of monolingual language models. https://doi.org/10.48550/ARXIV.2112.06598, https://arxiv.org/abs/2112.06598

Moon S, Okazaki N (2020) PatchBERT: Just-in-time, out-of-vocabulary patching. In: Proceedings of the 2020 Conference on Empirical Methods in Natural Language Processing (EMNLP). Association for Computational Linguistics, pp 7846–7852, Online. https://doi.org/10.18653/v1/2020.emnlp-main.631, https://aclanthology.org/2020.emnlp-main.631

Ruzzetti ES, Ranaldi L, Mastromattei M, Fallucchi F, Zanzotto FM (2021) Lacking the embedding of a word? look it up into a traditional dictionary. https://doi.org/10.48550/ARXIV.2109.11763, https://arxiv.org/abs/2109.11763

Sachidananda V, Kessler J, Lai YA (2021) Efficient domain adaptation of language models via adaptive tokenization. In: Proceedings of the Second Workshop on Simple and Efficient Natural Language Processing. Association for Computational Linguistics, Virtual, pp 155–165. https://doi.org/10.18653/v1/2021.sustainlp-1.16, https://aclanthology.org/2021.sustainlp-1.16

Tai W, Kung HT, Dong X, Comiter M, Kuo CF (2020) exBERT: Extending pre-trained models with domain-specific vocabulary under constrained training resources. In: Findings of the Association for Computational Linguistics: EMNLP 2020. Association for Computational Linguistics, pp 1433–1439, Online. https://doi.org/10.18653/v1/2020.findings-emnlp.129, https://aclanthology.org/2020.findings-emnlp.129

Wang H, Yu D, Sun K, Chen J, Yu D (2019) Improving pre-trained multilingual model with vocabulary expansion. In: Proceedings of the 23rd Conference on Computational Natural Language Learning (CoNLL). Association for Computational Linguistics, Hong Kong, China, pp 316–327. https://doi.org/10.18653/v1/K19-1030, https://aclanthology.org/K19-1030

Wang Z, K K, Mayhew S, Roth D (2020) Extending multilingual BERT to low-resource languages. In: Findings of the Association for Computational Linguistics: EMNLP 2020. Association for Computational Linguistics, pp 2649–2656, Online. https://doi.org/10.18653/v1/2020.findings-emnlp.240, https://aclanthology.org/2020.findings-emnlp.240

Yu W, Zhu C, Fang Y, Yu D, Wang S, Xu Y, Zeng M, Jiang M (2021) Dict-BERT: Enhancing language model pre-training with dictionary. https://doi.org/10.48550/ARXIV.2110.06490, https://arxiv.org/abs/2110.06490

Part II
Event-Centric Multilingual Knowledge Technologies

In today's digital age, the Web and social media are crucial for accessing news and information, and they cover a large number of events occurring worldwide on a daily basis. These events shape our perception of the world and can have substantial influence on individuals, communities and society at large. Novel technologies that combine event-centric knowledge graphs with state-of-the-art approaches from artificial intelligence (AI) are of the utmost importance since they provide a structured framework to capture, connect and comprehend event-centric information. This opens up new possibilities for efficient exploration and analysis of historical and contemporary events, presenting many applications in fields such as information retrieval and digital humanities. The combination of cutting-edge AI approaches and knowledge graphs also provides a valuable source of accurate factual information that offers the potential to improve the reliability of new technologies, including generative AI models.

However, many challenges exist in creating event-centric knowledge technologies. For example, coping with the sheer quantity, diversity and complexity of events poses a significant hurdle. Although cross-domain knowledge graphs such as *Wikidata* and *DBpedia* encompass a wide spectrum of entities including events, they are incomplete, lacking important events, quotes or details. Media outlets might also cover the same event across various languages from different perspectives. Thus, an in-depth analysis of events requires novel AI solutions and applications, such as recommendation or question answering systems, based on multilingual, event-centric knowledge graph information. Access to structured information in a knowledge graph typically requires a profound understanding of its underlying schema and the *SPARQL* query language. Hence, innovative approaches including generative AI models are required to provide a more intuitive access to knowledge graph information, for instance, through conversational question answering systems.

This part covers four chapters that introduce novel AI technologies based on knowledge graphs to address the aforementioned challenges. Chapter 6 presents the multilingual, event-centric knowledge graph *EventKG*, as well as its extension, the *Open Event Knowledge Graph* (*OEKG*), which covers more than 1.7 million events in 15 languages. *EventKG* combines information collected from various heterogeneous data sources including *Wikipedia* and cross-domain knowledge graphs such as *Wikidata* and *DBpedia*. The *OEKG* integrates several cross-lingual, event-centric datasets introduced throughout this book to extend the *EventKG*. Quotes are important parts of events and provide insights into the actions and opinions of public figures. Chapter 7 presents a multilingual knowledge graph called *QuoteKG* that covers nearly one million quotes from 69,000 people of public interest in 55 languages. A cross-lingual language model is introduced that aligns quotes across different languages and events, which allows for many applications including the creation of quote collection for particular events and the event-centric analysis of quotes. To analyse events across different languages, Chap. 8 suggests an entity recommendation system that learns spatio-temporal and language-specific features of events in a learning-to-rank model. For this purpose, the model integrates event-centric user interactions in different languages from *EventKG* and *Wikipedia* clickstreams collected in a new dataset called *EventKG+Click*. In this way, the system can take into account cultural and linguistic differences for a more accurate, language-specific event recommendation. Finally, Chap. 9 proposes two approaches for conversational question answering over knowledge graphs. They focus on general-purpose question answering and provide an important foundation to train models for event-centric, multilingual information that allow users to ask questions in natural language to facilitate access to knowledge graph information and receive relevant answers about events.

Eric Müller-Budack

Chapter 6
Collection and Integration of Event-Centric Information in Cross-Lingual Knowledge Graphs

Simon Gottschalk

Abstract Collecting and integrating event information in a knowledge graph enables the analysis of major societal events, their interdependencies with other events and actors and their perception and impact. While existing cross-domain knowledge graphs such as Wikidata and DBpedia also contain event knowledge, they are typically limited regarding the diversity of event representations and types. In this chapter, we first describe *EventKG*—a knowledge graph of multilingual event-centric information bringing together heterogeneous event information from different sources. Since the thorough understanding of events further demands the availability of context information in different modalities, we then present the Open Event Knowledge Graph (*OEKG*), which extends the coverage and modality of *EventKG* by integrating several of the event-related datasets presented in this book and opens up several possibilities for cross-lingual, event-centric open analytics. Through several statistics, example queries and applications, we show the versatility and the applicability of *EventKG* and *OEKG* for event analytics across languages and communities.

6.1 Introduction

The amount of event-centric information regarding contemporary and historical events of global importance, such as the US presidential elections and the Coronavirus pandemic, constantly grows on the Web, in news sources and on social media. Efficiently accessing and analysing large-scale event-centric and temporal information is crucial for various real-world applications in semantic Web, natural language processing and digital humanities.

Event knowledge is already captured in cross-domain knowledge graphs like Wikidata (Vrandečić and Krötzsch 2014) and DBpedia (Auer et al. 2007). However,

---

S. Gottschalk (✉)
L3S Research Center, Leibniz Universität Hannover, Hannover, Germany
e-mail: gottschalk@L3S.de

I. Marenzi et al. (eds.), *Event Analytics across Languages and Communities*,
https://doi.org/10.1007/978-3-031-64451-1_6

through their diverse focus, cross-domain knowledge graphs typically lack events not explicitly named (e.g. company foundations) or not considered relevant enough (e.g. politicians' speeches). On the other hand, event-centric datasets such as GDELT (Leetaru and Schrodt 2013) and Event Registry (Leban et al. 2014) typically cover very fine-grained events, often extracted from single sentences in news articles, that are rarely classified into semantic classes and only insufficiently interlinked with other entities in knowledge graphs. We extract relevant event information from several sources aiming at a broad coverage of events and their extensive representation, and we aim to further enrich event knowledge with more contextual information and modalities.

In this chapter, we first present the newest version of *EventKG*—a multilingual knowledge graph that contains event-centric information in the form of well-defined *named events*, *text events* described through short texts and *temporal relations*. *EventKG* integrates information from Wikipedia and cross-domain knowledge graphs. With a coverage of 15 languages, 1.7 million events and more than 50 million relations, *EventKG* can serve as a substantial knowledge source to conduct analyses of events across languages.

We also describe the Open Event Knowledge Graph (*OEKG*) first presented in Gottschalk et al. (2021)—an extension of *EventKG* that makes a step toward a holistic representation of event knowledge by integrating event-related datasets from multiple, diverse application domains such as question answering, entity recommendation and named entity recognition. These datasets were extracted from datasets such as news articles and image collections and are listed in Table 6.1. They are further described in other chapters of this book (as listed in the table).

At the examples of *EventKG* and *OEKG*, we show the value of event-centric knowledge graphs for event analytics across languages and communities. For example, both knowledge graphs can be used for tasks like event-centric question answering and language-specific recommendations and as a background source for several visual tools and user interfaces that allow the in-depth analysis of a single event and the exploration of multiple events, e.g. in a person's lifetime.

The remainder of this chapter is structured as follows: first, we present *EventKG* and *OEKG* in Sects. 6.2 and 6.3; Sect. 6.4 shows example queries and applications of these two knowledge graphs; and finally, we conclude in Sect. 6.5.

6.2 *EventKG*

Events of societal importance come in many forms, including:

- *named events* such as the Wimbledon Championships 2023
- *text events* such as "Louis VIII becomes King of France upon the death of his father, Philip II."
- *temporal relations* such as the marriage between Brad Pitt and Angelina Jolie from 2014 to 2019

Table 6.1 Statistics of the different datasets contained in the *OEKG*

Dataset	In this book	Short description	Triples
EventKG$_{light}$ (Gottschalk and Demidova 2019)	Sect. 6.2	A light-weight version of *EventKG* V3.0[a]	434,752,387
UNER (Alves et al. 2020)	Chap. 1	The universal named entity recognition framework	206,622
MLM (Armitage et al. 2020)	Chap. 2	A benchmark dataset for multitask learning with multiple languages and modalities	942,753
EventKG+Click (Abdollahi et al. 2020)	Chap. 8	A dataset of language-specific event-centric user interaction traces	118,662
VQuAnDa (Kacupaj et al. 2020)	Chap. 9	A verbalisation question answering dataset	38,243
InfoSpread (Sittar et al. 2020)	Chap. 10	A dataset for information spreading over the news	277,992
TIME (Cheema et al. 2020)	–	Two collections of news articles related to the Olympic legacy and Euroscepticism	70,754
OEKG		The Open Event Knowledge Graph	436,407,413

[a] *EventKG*$_{light}$ has no link and co-mention relations and provides less provenance information

To integrate all such forms of events into a common resource and thus allow rich event analytics, we have created *EventKG*. In the following, we explain its creation (Sect. 6.2.1) and its schema (Sect. 6.2.2) and provide statistics, specifically about its multilingual coverage (Sect. 6.2.3).

6.2.1 Creation

EventKG, first presented in Gottschalk and Demidova (2018a) and available online,[1] is a multilingual and event-centric knowledge graph with named events, text events and temporal relations. To gain such information, *EventKG* integrates data from the following sources:

- The cross-domain knowledge graphs Wikidata (Vrandečić and Krötzsch 2014), DBpedia (Auer et al. 2007) and YAGO (Suchanek et al. 2007): Events, entities, their attributes (e.g. labels and dates), (temporal) relations and types.
- Wikipedia:
  - Articles: Event descriptions, link relations and co-mention relations.

[1] https://eventkg.l3s.uni-hannover.de/.

- Event lists:[2] Text events.
- Wikipedia Current Events Portal:[3] Text events.

EventKG provides the original information from these sources in respective named graphs for provenance.[4] However, during a fusion step, selected triples regarding event dates, locations and coordinates are integrated and made available in an additional, integrated named graph.

6.2.2 Schema

The goals of the *EventKG* data model are to facilitate a lightweight integration and fusion of heterogeneous event representations and temporal relations extracted from the reference sources and make this information available through an RDF representation.

In *EventKG*, we build upon the simple event model (SEM) (Van Hage et al. 2011) as a basis to model events in RDF. SEM is a flexible data model that provides a generic event-centric framework. In addition to SEM, within the *EventKG* schema, we adopt additional properties and classes to adequately represent the information extracted from the reference sources, model temporal and event relations and provide provenance information. The schema of *EventKG* is presented in Fig. 6.1.[5]

6.2.3 Statistics

In March 2023, we published version 3.2 of EventKG, with the most recent data available from its sources. Table 6.2 gives an overview of selected RDF classes in *EventKG* and how often they are used. For instance, *EventKG* V3.2 has nearly one million named events plus approximately 700,000 text events.

[2] Example event list in the English Wikipedia: https://en.wikipedia.org/wiki/1967_in_music.

[3] Example month: https://en.wikipedia.org/wiki/Portal:Current_events/January_2016.

[4] For example, there is one named graph for triples extracted from Wikidata, one named graph for each Wikipedia language version, etc.

[5] The following prefixes and namespaces are used in this chapter: `dbo` (http://dbpedia.org/ontology/), `dcterms` (http://purl.org/dc/terms/), `eventkg-g` (https://eventkg.l3s.uni-hannover.de/graph/), `eventkg-s` (https://eventkg.l3s.uni-hannover.de/schema/), `owl` (http://www.w3.org/2002/07/owl#), `rdf` (http://www.w3.org/1999/02/22-rdf-syntax-ns#), `sem` (http://semanticweb.cs.vu.nl/2009/11/sem/), `skos` (http://www.w3.org/2004/02/skos/core#), `so` (http://schema.org/), `time` (http://www.w3.org/2006/time#), `uner` (http://oekg.l3s.uni-hannover.de/uner/).

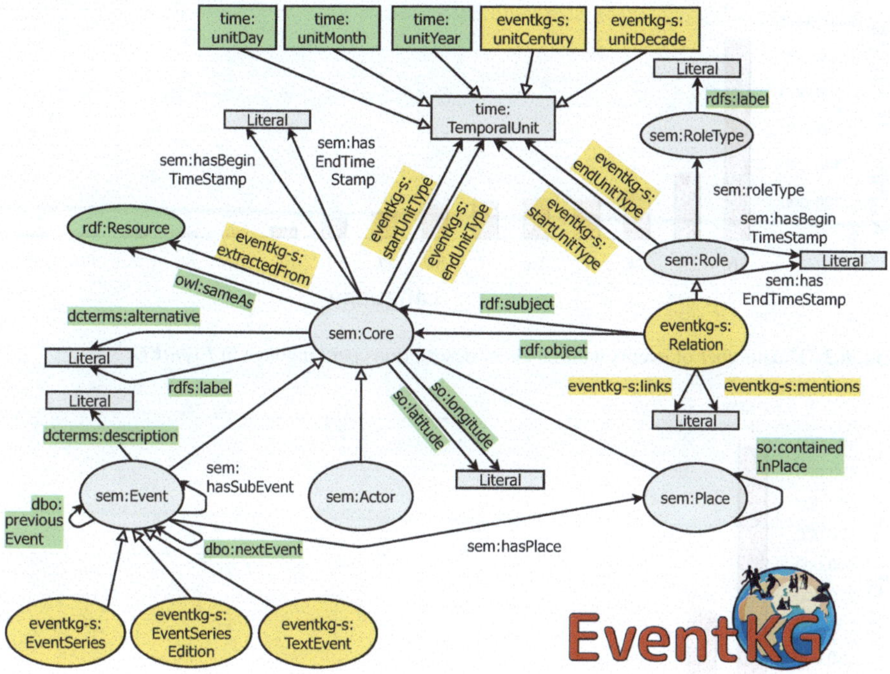

Fig. 6.1 The *EventKG* schema based on SEM. Arrows with an open head denote `rdfs:subClassOf` properties. Regular arrows visualise the `rdfs:domain` and `rdfs:range` restrictions on properties. Terms from other reused vocabularies are coloured green. Classes and properties introduced in *EventKG* are coloured orange

Table 6.2 Statistics of how often selected classes are used in *EventKG*. For brevity and intuition, the example column does not show the exact triples and URIs as they are given in *EventKG*

Class	Count	Example
`eventkg-s:Relation`	51,918,828	*Wimbledon Championships 2022, champion in single male, Novak Djokovic*
`eventkg-s:LinkRelation`	48,846,603	*The French Wikipedia article of Wimbledon Championships 2022 mentions Novak Djokovic 6 times.*
`eventkg-s:CoMentionRelation`	40,159,072	*There are 3 sentences in the French Wikipedia mentioning the Wimbledon Championships 2022 and Novak Djokovic.*
`dbo:Person`	5,967,835	Novak Djokovic
`sem:Place`	1,978,991	London
`sem:Event`	993,268	Wimbledon Championships 2023
`eventkg-s:TextEvent`	709,689	Wimbledon announced that the prize money will be raised by 40% in the 2013 tournament.
`eventkg-s:EventSeriesEdition`	201,468	Wimbledon Championships 2023
`eventkg-s:EventSeries`	97,085	Wimbledon Championships

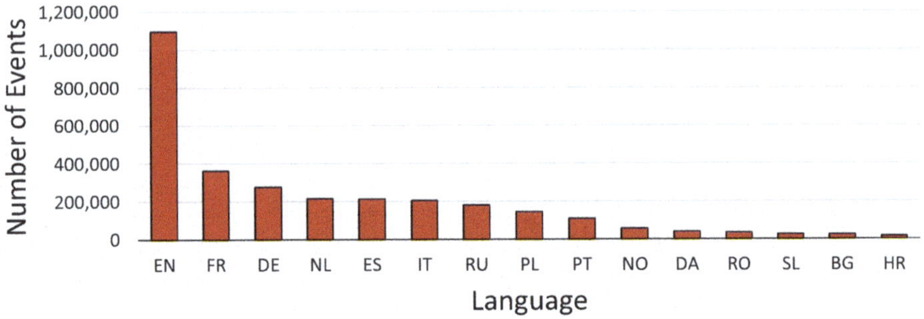

Fig. 6.2 The number of events with labels or descriptions per language in *EventKG*

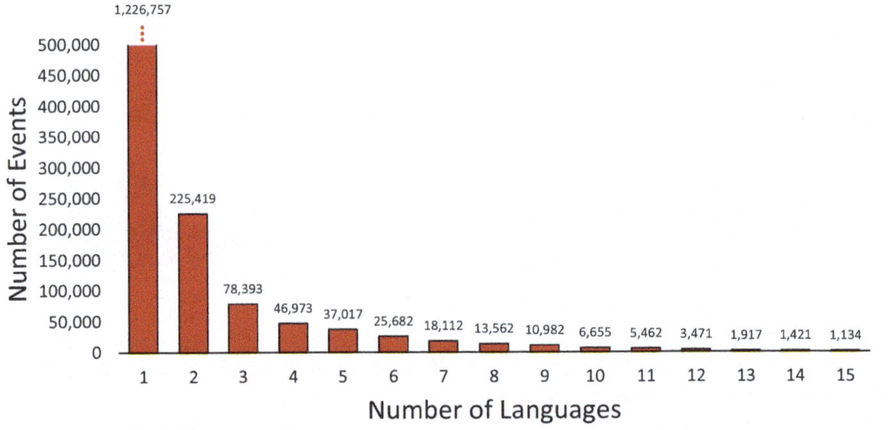

Fig. 6.3 The number of events that have a label or description in exactly 1 to 15 languages

Language Coverage in *EventKG*

In version 3.2, *EventKG* integrates information in 15 languages.[6] Figure 6.2 gives an overview of the event coverage per language: more than one million events in *EventKG* have an English label (in case of `sem:Event`) or an English description (in case of `eventkg-s:TextEvent`) available. Even for lower-resourced languages such as Norwegian (56,446 events) and Croatian (16,485), tens of thousands of events provide such language-specific information.

Another question regarding language coverage is how well events are aligned across languages. Figure 6.3 shows the number of events with labels or descriptions available in multiple languages. While approximately 72% of events are labelled in only one language (e.g. "2020 Illinois Democratic presidential primary" (EN)

[6] EN English, FR French, DE German, NL Dutch, ES Spanish, IT Italian, RU Russian, PL Polish, PT Portuguese, NO Norwegian, DA Danish, RO Romanian, SL Slovene, BG Bulgarian and HR Croatian.

and "Osmanski napad na Split" (HR)), nearly half a million events are labelled in 2 up to 15 languages. Examples of text events with a description in each of the 15 languages are "With the papal bull Inter gravissimas, Pope Gregory XIII announces the Gregorian calendar" and "Queen Victoria succeeds to the British throne".

6.3 *OEKG*

The Open Event Knowledge Graph (*OEKG*), first presented in Gottschalk et al. (2021) and available online,[7] is an extension of *EventKG* with six datasets from multiple application domains, including question answering, entity recommendation and named entity recognition as listed in Table 6.1. All these datasets were created as part of the CLEOPATRA ITN project introduced in this book's introduction and are briefly described in the following.

- ***EventKG$_{light}$*** (Gottschalk and Demidova 2019) (Sect. 6.2): *EventKG$_{light}$* is a lightweight version of EventKG that omits provenance information denoting the origin of relations, favouring an easier integration with other datasets. In the *OEKG*, *EventKG$_{light}$* serves as the base graph that other datasets are connected to.
- ***UNER*** (Alves et al. 2020) (Chap. 1): The Universal Named Entity Recognition framework proposes a four-level class hierarchy for training and testing named entity recognition tools (e.g. `Earthquake`). In the *OEKG*, *UNER* adds to the class hierarchy already provided by the DBpedia and Wikidata ontologies.
- ***MLM*** (Armitage et al. 2020) (Chap. 2): The Multiple Languages and Modalities dataset is a resource for training and evaluating multitask systems in multiple modalities, for example, cross-modal (text/image) retrieval and location estimation. *MLM* is added to the *OEKG* for adding images as an additional modality to the knowledge graph.
- ***EventKG+Click*** (Abdollahi et al. 2020) (Chap. 8): *EventKG+Click* is a cross-lingual dataset that reflects the language-specific relevance of events and their relations derived from the Wikipedia clickstream. In the *OEKG*, *EventKG+Click* can be used for recommending events to users based on actual user interaction traces.
- ***VQuAnDa*** (Kacupaj et al. 2020) (Chap. 9): The Verbalization QUestion ANswering DAtaset is a dataset for question-answering (QA) over knowledge graphs that includes the verbalisation of each answer. Via the integration of question/answer pairs into the *OEKG*, both the question/answer pairs and the background knowledge are encapsulated into the same resource, enabling seamless training and application of QA systems.
- ***InfoSpread*** (Sittar et al. 2020) (Chap. 10): The dataset for Information Spreading over the News provides news articles covering three contrasting events (Global

[7] https://oekg.l3s.uni-hannover.de/.

Warming, FIFA world cups and earthquakes). The inclusion of news articles into the *OEKG* is an important step towards coverage of event-centric data from different viewpoints.
- ***TIME*** (Cheema et al. 2020): The temporal discourse analysis applied to media articles dataset is a collection of Brazilian, British and Spanish news articles covering the concept of Olympic legacy and the concept of Euroscepticism. With the collection of news articles related to specified events, the *OEKG* serves as an example for in-depth analysis of single events through knowledge graphs.

6.3.1 Creation

To integrate several heterogeneous datasets into one knowledge graph, we build an *OEKG* integration pipeline that follows a strategy defined by Galkin et al. (2016), where the data from different sources is stored under the respective named graphs. As shown earlier, seven datasets listed in Table 6.1 are integrated into the *OEKG*, including the number of triples in the *OEKG* within their respective named graph.

6.3.2 Schema

The *OEKG* schema is based on the *EventKG* schema (Fig. 6.1) and was extended following the demand of the respective datasets. For example, a question, its suggested answer and verbalisation from *VQuAnDa* (Kacupaj et al. 2020) (Chap. 9) are represented using `schema.org`'s classes `so:Question` and `so:Answer`. Images from *MLM* (Armitage et al. 2020) (Chap. 2) are assigned to places via `so:image`, descriptions via `so:description`.

6.4 Applications

In this section, we provide example queries for *EventKG* and *OEKG*, showcasing how to extract structured event information from both sources.

6.4.1 Example Query on EventKG

To explore an event to its full extent, one needs to know not only the attributes, such as its date and descriptions, but also its sub-events, i.e. events that it is composed of. Listing 6.1 is a SPARQL query for *EventKG* that retrieves the sub-events of the French presidential election in 2022 (via `sem:hasSubEvent`) and their start dates (via `sem:hasBeginTimeStamp`).

```
SELECT ?StartDate ?Label WHERE {
  ?event owl:sameAs dbr:2022_French_presidential_election .
  GRAPH eventkg-g:event_kg {
    ?subEvent sem:hasBeginTimeStamp ?StartDate .
  }
  ?event sem:hasSubEvent ?subEvent .
  ?subEvent skos:prefLabel ?Label .
} ORDER BY ?StartTime
```

Listing 6.1 SPARQL query for *EventKG* to retrieve the sub-events of the French presidential election in 2022 and their start dates

Table 6.3 *EventKG* results of the SPARQL query in Listing 6.1

?StartDate	?Label
2021-05-01	"Consultation interne au Parti communiste français pour l'élection présidentielle de 2022"@fr
2022-04-10	"First round of French presidential election, 2022"@en
2022-04-24	"Second round of French presidential election, 2022"@en

Table 6.3 shows the retrieved results for the query, including the two rounds of the election, which are indispensable to consider when analysing the French presidential election in 2022.

6.4.2 *Example Query on the* OEKG

The *OEKG* facilitates queries for the *UNER* type hierarchy specifically designed for named entity recognition and for images of locations, using the *MLM* data. In combination, event locations in $EventKG_{light}$, *MLM*'s image links and the *UNER* type hierarchy enable retrieval of images relevant to specific event types.

We demonstrate the *OEKG*'s potential for image retrieval with an example query for images from earthquake regions shown in Listing 6.2: it queries entities typed as earthquakes using the newly created `uner:Earthquake` class, their locations ($EventKG_{light}$) and the images assigned to such locations (*MLM*). Amongst others, the results of this query include three cities (`?Location`: Ferrara, Messina and Guaranda), each together with a photo from Wikimedia (`?Image`).

```
SELECT DISTINCT ?Location ?Image WHERE {
  ?earthquake rdf:type uner:Earthquake ;
    sem:hasPlace ?Location  .
  ?Location so:image ?Image .
}
```

Listing 6.2 SPARQL query for the *OEKG* to retrieve images of locations where earthquakes happened

6.4.3 Example Applications of EventKG

EventKG has been used in different applications and datasets, showing its versatility and the applicability of event-centric knowledge graphs for event analytics across languages and communities.

- Visualisations and user interfaces: We have developed several tools to visualise the information inherent in *EventKG*:
  - *EventKG+BT* (Gottschalk and Demidova 2019, 2020) enables a user to explore the lives of any persons in *EventKG*. Instead of making users read a whole biography text about a person of interest, they can easily interact with the generated timeline and follow what was important in that person's life.
  - *EventKG+TL* (Gottschalk and Demidova 2018b) allows a user to explore a topic of interest by showing related events and their relevance from the points of view of different languages.
  - *VisKonnect* (Latif et al. 2021) visualises the connections between historical figures in the context of events, combining a chat interface using the GPT-3 language model and various linked visualisations.

  The described tools take a step toward reducing the workload that is necessary when closely reading encyclopaedic articles or exploring events of interest.
- Event-centric question answering: *EventQA* (Souza Costa et al. 2020) is a question-answering dataset specifically designed to evaluate the capability of question-answering models to answer questions regarding temporal and event-centric information. Examples of the 1000 queries available in English, German and Portuguese are "Give me a list of football games won by Dynamo Kyiv" and "When did the Excitante music festival finish in Argentina?".
- Selected information in *EventKG* and *OEKG* has been extracted for several other tasks, including:
  - Creation of the Visual Event Ontology for image classification (Müller-Budack et al. 2021).
  - Anomaly detection to identify the major events of public figures (Guo et al. 2022).
  - Improvement of annotations for named entity recognition as discussed in Chap. 6.
  - Language-specific event recommendations as discussed in Chap. 8.

6.5 Conclusion and Future Work

In this chapter, we presented *EventKG*—a multilingual knowledge graph incorporating event-centric information—and provided statistics of its newest version regarding its coverage of 15 languages. With more than 1.7 million events that are represented in a common schema and interlinked and a broad coverage of different event types, *EventKG* has proven to be a rich resource of event-centric knowledge. The Open Event Knowledge Graph (*OEKG*) is an extension of *EventKG* covering additional facets of event knowledge, such as a question-answering resource and image data.

We demonstrate the use of *EventKG* and *OEKG* for event analytics across languages through selected queries and present a set of existing applications and datasets built on top of them. Example applications include named entity recognition, language-specific event recommendations and the narrativisation of events, all discussed in other chapters of this book. In the future, we plan to further extend our event knowledge graphs, e.g. through frequent updates and the integration of further datasets such as *QuoteKG* (Chap. 7).

Acknowledgments The work presented in this chapter has received funding from the European Union's Horizon 2020 research and innovation program under the Marie Skłodowska-Curie grant agreement no. 812997 and under the name CLEOPATRA (Cross-lingual Event-centric Open Analytics Research Academy)

References

Abdollahi S, Gottschalk S, Demidova E (2020) EventKG+Click: A dataset of language-specific event-centric user interaction traces. In: Proceedings of the CLEOPATRA Workshop at the 19th International Semantic Conference

Alves D, Kuculo T, Amaral G, Thakkar G, Tadic M (2020) UNER: Universal named-entity recognition framework. In: Proceedings of the CLEOPATRA Workshop at the 19th International Semantic Conference

Armitage J, Kacupaj E, Tahmasebzadeh G, Maleshkova M, Ewerth R, Lehmann J (2020) MLM: A benchmark dataset for multitask learning with multiple languages and modalities. In: Proceedings of the 29th ACM International Conference on Information & Knowledge Management (CIKM), pp 2967–2974

Auer S, Bizer C, Kobilarov G, Lehmann J, Cyganiak R, Ives ZG (2007) DBpedia: A nucleus for a web of open data. In: The Semantic Web, 6th International Semantic Web Conference, 2nd Asian Semantic Web Conference, ISWC 2007. Lecture Notes in Computer Science, vol 4825. Springer, pp 722–735

Cheema GS, Major D, Mello C, Sittar A (2020) TIME: Temporal discourse analysis applied to media articles. http://cleopatra-project.eu/index.php/2020/06/01/time-temporal-discourse-analysis-applied-to-media-articles/

Galkin M, Auer S, Kim H, Scerri S (2016) Integration strategies for enterprise knowledge graphs. In: Proceedings of the IEEE Tenth International Conference on Semantic Computing (ICSC), pp 242–245

Gottschalk S, Demidova E (2018a) EventKG: A multilingual event-centric temporal knowledge graph. In: The Semantic Web: 15th International Conference (ESWC). Springer, pp 272–287

Gottschalk S, Demidova E (2018b) EventKG+TL: Creating cross-lingual timelines from an event-centric knowledge graph. In: The Semantic Web: ESWC 2018 Satellite Events. Springer, pp 164–169

Gottschalk S, Demidova E (2019) EventKG—The hub of event knowledge on the web—and biographical timeline generation. Semantic Web 10(6):1039–1070

Gottschalk S, Demidova E (2020) EventKG+BT: Generation of interactive biography timelines from a knowledge graph. In: The Semantic Web: ESWC 2020 Satellite Events. Springer, pp 91–97

Gottschalk S, Kacupaj E, Abdollahi S, Alves D, Amaral G, Koutsiana E, Kuculo T, Major D, Mello C, Cheema GS, Sittar A, Swati, Tahmasebzadeh G, Thakkar G (2021) OEKG: The open event knowledge graph. In: Proceedings of the CLEOPATRA Workshop at the 30th The Web Conference

Guo X, Zhou B, Skiena S (2022) Subset node anomaly tracking over large dynamic graphs. In: Proceedings of the 28th ACM SIGKDD Conference on Knowledge Discovery and Data Mining (KDD), pp 475–485

Kacupaj E, Zafar H, Lehmann J, Maleshkova M (2020) VQuAnDa: Verbalization QUestion ANswering DAtaset. In: The Semantic Web - 17th International Conference (ESWC). Springer, pp 531–547

Latif S, Agarwal S, Gottschalk S, Chrosch C, Feit F, Jahn J, Braun T, Tchenko YC, Demidova E, Beck F (2021) Visually connecting historical figures through event knowledge graphs. In: Proceedings of the 2021 IEEE Visualization Conference (VIS). IEEE, pp 156–160

Leban G, Fortuna B, Brank J, Grobelnik M (2014) Event registry: Learning about world events from news. In: Proceedings of the 23rd International Conference on World Wide Web (WWW), pp 107–110

Leetaru K, Schrodt PA (2013) GDELT : Global data on events, location, and tone, 1979–2012. In: ISA Annual Convention. Citeseer, pp 1–49

Müller-Budack E, Springstein M, Hakimov S, Mrutzek K, Ewerth R (2021) Ontology-driven event type classification in images. In: Proceedings of the IEEE Winter Conference on Applications of Computer Vision (WACV), pp 2928–2938

Sittar A, Mladenić D, Erjavec T (2020) A dataset for information spreading over the news. In: Proceedings of the Slovenian KDD Conference on Data Mining and Data Warehouses (SiKDD)

Souza Costa T, Gottschalk S, Demidova E (2020) Event-QA: A dataset for event-centric question answering over knowledge graphs. In: Proceedings of the 29th ACM International Conference on Information & Knowledge Management (CIKM), pp 3157–3164

Suchanek FM, Kasneci G, Weikum G (2007) YAGO: A core of semantic knowledge. In: Proceedings of the 16th International Conference on World Wide Web (WWW), pp 697–706

Van Hage WR, Malaisé V, Segers R, Hollink L, Schreiber G (2011) Design and use of the simple event model (SEM). Web Semantics 9(2):128–136. https://doi.org/10.1016/j.websem.2011.03.003

Vrandečić D, Krötzsch M (2014) Wikidata: A free collaborative knowledgebase. Commun ACM 57(10):78–85

Chapter 7
Event Analysis Through *QuoteKG*: A Multilingual Knowledge Graph of Quotes

Tin Kuculo and Simon Gottschalk

Abstract Significant moments in history are often remarked upon by public figures in the form of quotes. As evidence of character traits and future political or personal decisions, quotes provide insight into the actions of their originators. The impact of a quote crosses language barriers and influences the public's reaction to specific political stances. Nevertheless, effectively collating, attributing and analysing these quotes across languages remain challenging. Existing efforts have made strides in quote collections and analyses, yet several limitations persist, including a lack of context information, a labour-intensive extraction process and missing alignment of quote mentions across languages. Building upon *QuoteKG*, a multilingual knowledge graph of quotes that already addresses some of the aforementioned limitations, we present an approach for aligning quotes with event knowledge. *QuoteKG* is based on *Wikiquote*, a free and collaboratively created collection of quotes in many languages. Containing nearly one million quotes in 55 languages said by 69,000 people of public interest, *QuoteKG* extracts and aligns different mentions and contexts of quotes across a wide range of topics. We show that *QuoteKG* can be aligned with event knowledge. We use this alignment to enrich and analyse event-centric information by providing rich semantic context to important world events. *QuoteKG* is publicly available and can be accessed via a *SPARQL* endpoint.

7.1 Introduction

Quotes from public figures provide valuable information to understand their thoughts and attitudes, potentially leading to historically important actions, and thus serve as a crucial component in exploring world history (Khurana 2018). Table 7.1 provides three examples of quotes, with the first one emphasising the

T. Kuculo (✉) · S. Gottschalk
L3S Research Center, Leibniz Universität Hannover, Hannover, Germany
e-mail: kuculo@L3S.de; gottschalk@L3S.de

I. Marenzi et al. (eds.), *Event Analytics across Languages and Communities*,
https://doi.org/10.1007/978-3-031-64451-1_7

Table 7.1 Three example quotes, together with their originators and dates. The last column gives examples of context that can be attributed to the mention of a quote, including source information, translations or validation of the quote's correctness

Quote	By	Date	Selected context
It is a good thing for an uneducated man to read books of quotations. (English)	Winston Churchill	1930	*Source:* Roving Commission: My Early Life (1930) Chap. 9
Wir schaffen das. (German)	Angela Merkel	2015, Aug 31	*Translation:* We can do this (English)
The definition of insanity is doing the same thing over and over and expecting different results. (English)	Albert Einstein		Misattributed

relevance of historical quotes: in 1930, Winston Churchill recognised the value of reading them. The second example in Table 7.1 illustrates the relevance of quotes in world history: during a press conference in 2015, the German chancellor Angela Merkel said, "Wir schaffen das" ("We can do this"), when the European migrant crisis unfolded, and Germany prepared for the reception of refugees from Northern Africa and the Middle East. Since then, these three words have defined Merkel's political course in the migrant crisis—and led both to a welcoming culture and the rise of nationalist protests and right-wing political parties (Mushaben 2017; Krämer 2021).

Given this potential impact of words, it is of the utmost importance to provide sources for quotes and to dismiss hoaxes (Keyes 2007; Robinson 2018): The third example in Table 7.1 is a famous quote that has been attributed to different people, including Albert Einstein, Benjamin Franklin and Mark Twain, but has not actually been said by any of them.[1] In general, a quote can be mentioned in different sources, and mentions can deviate. For example, "Wir schaffen das" might be mentioned as "We can do this" or "We will make it" in English translations. Therefore, there is a need to align mentions to the same quote and to provide context information such as the source and description (e.g. "during a press conference").

7.1.1 Challenges

The creation of a knowledge graph covering quotes in many languages and their contexts poses several challenges. In current works, Newell et al. (2018) focus on searchable databases from journalistic sources, analysing quote usage and topics. Goel et al. (2018) devise a system for formulating quote-like captions for images by combining deep learning and natural language processing. Meanwhile, Vaucher

[1] Reasons for false attribution of quotes to persons include to appear educated or to lend authority from the person (Reucher 2021).

et al. (2021) present *Quotebank* and *Quobert*, tools for extracting and attributing quotations from extensive corpora. However, there are limitations to these existing works, which our approach aims to address:

- *Lack of context*: Most quote collections (Newell et al. 2018; Goel et al. 2018; Vaucher et al. 2021) lack context information and solely provide the quotes and their originators. To provide more context information in *QuoteKG*, we extract quotes from *Wikiquote*—a "free online compendium of sourced quotes from notable people".[2]
- *Societal relevance of quotes*: Many existing corpora focus on quote extraction primarily from news articles. Although this provides a large volume of quotes, it may not always ensure the societal relevance of the quotes. In contrast, *Wikiquote*, from which *QuoteKG* extracts quotes, is known for its curated collection of societally relevant quotes from prominent individuals across multiple languages and cultures.
- *Tedious extraction process*: Extraction of quotes and contexts from semi-structured resources like *Wikiquote* is a tedious and complex process. In particular, we must design an extraction pipeline that is flexible across languages and adopts their characteristics. For example, it is necessary to differentiate the quotes not said by a person but said about a person (e.g. English, "Quotes about Albert Einstein"; German, "Zitate mit Bezug auf Albert Einstein").
- *Missing alignment of quote mentions:* As quote mentions in *Wikiquote* are not linked across languages, another important step is cross-lingual quote alignment, which we perform using a language-agnostic transformer model that we evaluate on a ground-truth set of manually aligned quote clusters.

7.1.2 Contributions

In this chapter, we detail our contributions, highlighting how they address the identified challenges:

(i) We present *QuoteKG*, a comprehensive knowledge graph consisting of around one million quotes expressed in 55 languages by more than 69,000 public figures that was first introduced by Kuculo et al. (2022). This knowledge graph incorporates sentiment and context with quotes, thereby tackling the lack of context prevalent in current quote collections. By focusing on quotations from notable public figures, *QuoteKG* guarantees a degree of societal relevance, offering a broader insight into global discourse and trends. (ii) We have developed a schema specifically designed to encapsulate both quotes and their context information, enhancing the understanding and usability of the data within our knowledge graph. (iii) Addressing the tedious extraction process, we have constructed a flexible pipeline capable of

[2] https://en.wikiquote.org/wiki/Main_Page

efficiently extracting quotes, their mentions and pertinent contextual information from *Wikiquote*. (iv) We have implemented a cross-lingual language model to align quote mentions across languages, addressing the issue of unlinked quote mentions. (v) We demonstrate the practical utility of *QuoteKG*, detailing its application in compiling event quote collections and performing event-centric quote analysis. (vi) As a commitment to open research and community contribution, we have made *QuoteKG* publicly available.[3]

7.1.3 *Event Analysis Through* QuoteKG

Quotes play an important role for event analytics across languages and communities discussed in this book. We specifically note the connections between our work and Chap. 11 about fact-checking and Chap. 6 on the Open Event Knowledge Graph (*OEKG*).

In Chap. 11, the crucial tasks of fact-checking and misinformation analysis are discussed extensively. Misinformation frequently comprises misattributed or inaccurately quoted statements, thereby underscoring the necessity of *QuoteKG*'s commitment to precise sourcing and attribution of quotes.

QuoteKG provides links between quotes and events in the *Wikidata* (Vrandecic and Krötzsch 2014) and *DBpedia* (Lehmann et al. 2015) knowledge graphs. Furthermore, in Sect. 7.6, we perform exemplary event analytics with *QuoteKG*, for instance, through the collection of event quote collections. A potential integration of such information with *OEKG* would further strengthen the access to semantic event information from knowledge graphs and foster a holistic understanding of events and their implications across diverse languages and communities.

7.1.4 *Outline*

The remainder of this chapter is structured as follows: First, in Sect. 7.2, we describe the schema adopted for *QuoteKG*. In Sect. 7.3, we describe the *QuoteKG* creation pipeline. In Sect. 7.4, we provide statistics and examples of *QuoteKG*, followed by information about the availability in Sect. 7.5. In Sect. 7.6, we use *QuoteKG* to build event quote collections and show examples of analyses that can be conducted on event-aligned quotes. Section 7.7 gives an overview of related work. Finally, we provide a conclusion in Sect. 7.8.

[3] https://quotekg.l3s.uni-hannover.de.

7.2 *QuoteKG* Schema

As a first step to creating *QuoteKG*, we introduce and define the *QuoteKG* schema. The goal of the *QuoteKG* schema is to model quotes and their relationships with persons and other entities, as well as their different mentions, e.g. translations, typically in different contexts. To this end, *QuoteKG* is based on an extension of the *schema.org* vocabulary that provides a `so:Quotation`[4] class, which is re-used. According to the schema.org description, the `so:Quotation` class models quotes that are "Often but not necessarily from some written work" and can also refer to a "Quotation from an Event".[5] Therefore, it fits well with our concept of a quote in *QuoteKG*. However, we extend the schema with a new class `qkg:Mention`, which models the different mentions of a quote.

Figure 7.1[6] presents *QuoteKG*'s schema. Its classes are described in the following:

- **Person**: Each quote in *QuoteKG* is assigned to a person modelled as `so:Person`. For persons, *QuoteKG* provides additional type information (e.g. Politician)

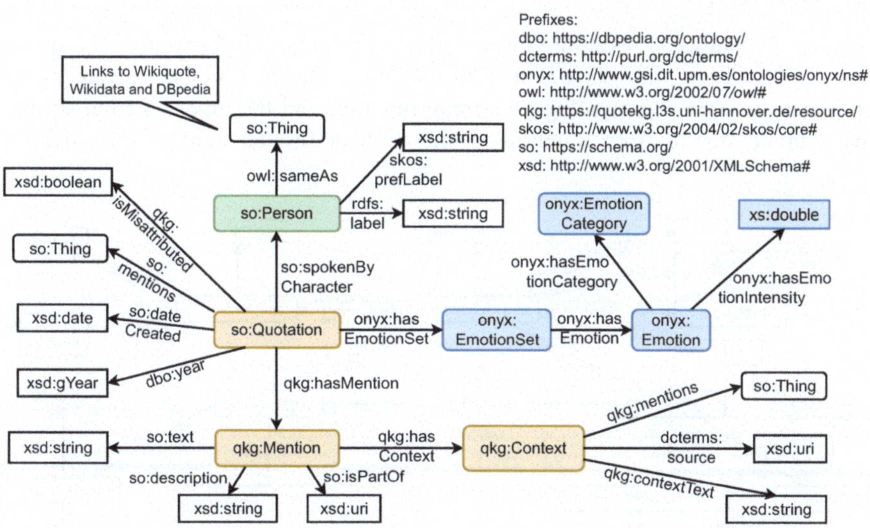

Fig. 7.1 The *QuoteKG* schema based on *schema.org*. Arrows visualise the `rdfs:domain` and `rdfs:range` restrictions on properties. Namespaces and prefixes are described in the top-right corner. Orange classes are related to quotes and their mentions and blue ones to the sentiment of a quote, and the green class is about the person

[4] So: https://schema.org/.

[5] https://schema.org/Quotation.

[6] Figures 7.1, 7.2, 7.3, 7.6 and 7.7 were created by Kuculo et al. (2022), published under CC BY 4.0 (https://creativecommons.org/licenses/by/4.0/) and have been updated to a new version for this book.

plus `owl:sameAs` relations to *Wikidata* (Vrandecic and Krötzsch 2014) and the different *DBpedia* (Auer et al. 2007) and *Wikiquote* language editions.

- **Quote**: In *QuoteKG*, a resource typed as `so:Quotation` refers to the unique event of something being said by a person of public interest (`so:spokenBy-Character`) at a specific point in time (`so:dateCreated`). A quote may also refer to other entities (`so:mentions`[7]) of any type, including events.
- **Mention**: A quote can be mentioned in different contexts: for example, there may be translations of the quote in different languages, alternative records of the same quote or different contexts that a quote is extracted from. Therefore, we introduce the class `qkg:Mention`. Mentions can be related to one or more `qkg:Context` objects.
- **Context**: The context of a mention provides additional attributes that come together with the specific mention, for example, its origin (e.g. a reference to a specific interview) and the original source (e.g. a link to a news Web site). To model context, we create the class `qkg:Context`.
- **Sentiment**: For each quote, we provide its sentiment using the Onyx ontology, which is used for describing emotions (Sánchez-Rada and Iglesias 2016). A quote is assigned a score for a specific emotion category ("neutral", "negative" or "positive").

Figure 7.2 shows an example instantiation of the *QuoteKG* schema. The quote "Wir schaffen das" introduced in Table 7.1 is connected to two instances of `qkg:Mention`, one representing a German mention and the other an English one ("We can do this"). Both mentions come with additional context information.

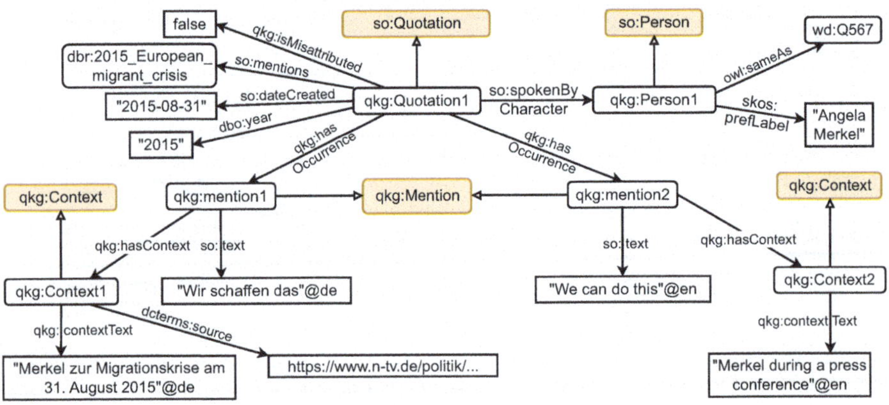

Fig. 7.2 An example quote modelled using the *QuoteKG* schema. ⇾ marks `rdf: type` relations. `xsd:` data type annotations were omitted for brevity. The prefixes and namespaces are the same as in Fig. 7.1, plus `wd` (https://www.wikidata.org/entity/) and `dbr` (https://dbpedia.org/resource/)

[7] `qkg:Mention` and `so:mentions` refer to different types of mentions.

We also show that the quote is linked to a mention of an event in *DBpedia* (`dbr: 2015_European_migrant_crisis`).

7.3 Extraction and Alignment of Quotes

This section describes the input data and the implementation of the four main steps of the *QuoteKG* creation pipeline shown in Fig. 7.3.

7.3.1 Wikiquote

We base *QuoteKG* on *Wikiquote*—an online collection of quotes.[2] *Wikiquote* has a similar structure to Wikipedia: independent versions of *Wikiquote* exist for different languages. *Wikiquote* contains pages, each of them about a given topic, divided into different sections and subsections. For *QuoteKG*, we focus on *Wikiquote* pages about persons that contain quotes attributed to them. Example pages are the English[8] and French[9] page about Albert Einstein.

Each *Wikiquote* page is formatted using the MediaWiki markup[10] and contains semi-structured content that includes the person's description, sections with quotes, references and more. The quotes are given in one of the following representations: in the traditional MediaWiki markup as shown in Fig. 7.4 or using pre-defined *templates* that allow for a more structured definition of key-value pairs. For example, Fig. 7.5 shows the key-value pair (*key*: `Citation`, *value*: `Tomber amoureux...`).

While there are links between the pages describing persons in different languages, quote mentions are not linked across languages. Figures 7.4 and 7.5 show

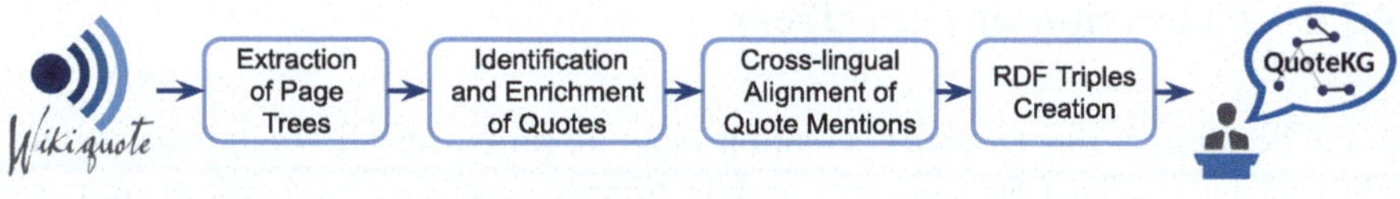

Fig. 7.3 Pipeline to create *QuoteKG* from *Wikiquote*

```
* '''Falling in love is not at all the most stupid thing that people do
- but gravitation cannot be held responsible for it.'''
** Jotted (in German) on the margins of a letter to him (1933), p. 56
```

Fig. 7.4 Example of a quote in the English *Wikiquote*, based on MediaWiki markup

[8] https://en.wikiquote.org/wiki/Albert_Einstein.

[9] https://fr.wikiquote.org/wiki/Albert_Einstein.

[10] https://www.mediawiki.org/wiki/Help:Formatting.

```
{{Citation|Tomber amoureux n'est pas du tout la chose la plus stupide
que font les gens - mais la gravitation ne peut en être tenue pour
responsable. |original=Falling in love is not at all the most stupid
thing that people do - but gravitation cannot be held responsible for
it.|langue=en}}
```

Fig. 7.5 Example of a quote in the French *Wikiquote*, using templates

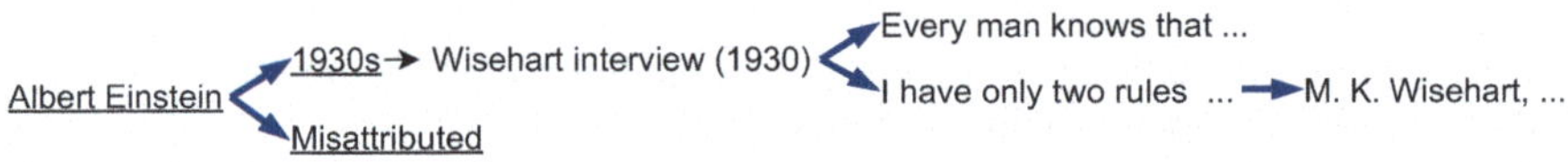

Fig. 7.6 Excerpt of an example page tree from the English *Wikiquote* page about Albert Einstein. Section titles are underlined

two mentions of the same quote by Albert Einstein. The first is from the English *Wikiquote* and shows an English quote, and the second one from the French *Wikiquote* is given in French and English. The original German quote is not available in these two language versions.

In general, one can observe a large imbalance in *Wikiquote* regarding the covered persons and the number of quotes in different language versions. This imbalance can often be explained by the different sizes of *Wikiquote* language versions and the difference in the cultural significance of a person in one language community compared to another. For example, there exists a French[11] page with 35 quotes and an Italian[12] page with 2 quotes from the former French footballer Michel Platini, who used to play in Italy, but there is no English page. This imbalance also implies that there is no guarantee that *Wikiquote* will contain the original language version of a quote. *QuoteKG* can have multiple quote mentions of the same quote through cross-lingual quote mention alignment.

7.3.2 Extraction of Page Trees

In the beginning, our *QuoteKG* creation pipeline processes all *Wikiquote* language editions with at least 50 pages, excluding Simple English,[13] and selects all pages about persons. From each *Wikiquote* page about a person, we create a *page tree*. The page tree consists of section titles plus quotes and contexts. An example page tree is presented in Fig. 7.6.

[11] https://fr.wikiquote.org/wiki/Michel_Platini.

[12] https://it.wikiquote.org/wiki/Michel_Platini.

[13] For more detailed statistics about *Wikiquote* language editions, see https://wikistats.wmcloud.org/display.php?t=wq.

7.3.3 Identification and Enrichment of Quotes

In the second step of the *QuoteKG* creation pipeline, page trees are transformed into a set of quotes with contextual information. To this end, we specify language-specific rules and enrich quotes and contexts with additional metadata.

To identify quotes, we first define a language-specific list of section titles denoting quotes (e.g. "Citations" in English, "Zitate" in German, "Citazioni" in Italian) and contextual information (e.g. "útskýring" in Icelandic, "Viitattu" in Finnish, "vydavatel" in Czech). In addition, we collect a list of template types representing quotes and consider all child nodes of section titles as quotes. From section titles, markup and templates, we further gather the following:

- Dates: We identify the dates of quotes from a pre-defined list of template keys (e.g. "année d'origine" in French) for quotes extracted from templates. If such dates are not available or when dealing with quotes not extracted from templates, we extract dates from the section titles above the particular quote in the page tree and the contexts below the quote.

  We select the time expression with the highest level of precision (e.g. we select May 2020 over 2020). In case of conflicts, no date is chosen.
- Veracity: To reflect the authenticity of quotes and their contextual information, we capture whether a quote has been misattributed to the person. In *Wikiquote*, misattributed quotes are grouped into specified sections. We identify such sections with a manually created list of regular expressions (e.g. "Misattributed" (English) and "Fälschlich zugeschrieben" (German).
- Sources: Often, context contains links to Web sites where the quote was reported. We collect such external links from templates and from the markup.
- Linked entities: Quotes can be linked to entities such as other persons, organisations or events. We collect such links from templates and markup.
- Language: While the *Wikiquote* pages are written in specific languages, their quotes can be written in their original language or translated. For this reason, we use an n-gram-based language detection tool[14] to designate the language of a quote and do not rely on the language of the page itself.
- Sentiment: We detect the sentiment of each quote mention (*positive*, *negative* or *neutral* with a score between 0 and 1) using XLM-RoBERTa-Twitter, an XLM-RoBERTa model trained on $\sim 198M$ multilingual tweets (Barbieri et al. 2022).
- Identity links: To establish `owl:sameAs` links between the *QuoteKG* entities, *Wikidata* and *DBpedia*, we use *Wikidata's* sitelinks.[15]

For all persons and entities identified during this process, we extract additional information regarding their labels and types from *DBpedia* and *Wikidata*.

[14] https://pypi.org/project/langdetect/.

[15] https://www.wikidata.org/wiki/Help:Sitelinks/en-gb.

7.3.4 Cross-lingual Alignment of Quote Mentions

After identifying and enriching quotes, we need to detect which of them represent mentions of the same quote said by a person of public interest. This task of cross-lingual alignment of quote mentions is treated as a clustering task, at the end of which each cluster represents a quote with a set of mentions.

In detail, the clustering task is performed for each person in isolation. Given a person's quote mentions in a set of languages, we aim at creating clusters of highly similar mentions based on a pre-defined similarity threshold parameter τ. This parameter is critical in the agglomerative clustering process, and its selection is detailed in Sect. 7.3.6. To derive a similarity between two mentions, potentially from different languages, we compute the cosine similarity of sentence embeddings derived from the mentions' texts. As an embedding model, we use a language-agnostic transformer model pre-trained on millions of multilingual paraphrase examples in more than 30 languages, namely, XLM-RoBERTa (Conneau et al. 2020). The ability of such models to adapt to previously unknown languages has been shown in Hu et al. (2020). Given these embeddings, the cosine similarity function and the aforementioned τ, clustering is performed by detecting communities of quotes using a nearest-neighbour search. To do so, we chose UKPLab's Fast Clustering algorithm,[16] optimised towards efficient similarity computations of our embeddings.

To aggregate the sentiments of all mentions in a cluster, we take the most frequent sentiment category and average over the scores of that category.

7.3.5 RDF Triples Creation

Following the identification of quotes and their contexts and the cross-lingual alignment, we transform the quotes into RDF triples following the schema presented in Fig. 7.1.

Listing 7.1 shows triples representing an example quote by the painter Arshile Gorky. The quote is linked to its speaker and the two persons it mentions, Pablo Picasso and Paul Cézanne. It is further contextualised with additional information such as the year it was said (1945) that it was not misattributed and its emotion (in this case neutral).

[16] https://github.com/UKPLab/sentence-transformers/blob/master/examples/applications/clustering/.

```
@prefix qkg: <https://quotekg.l3s.uni-hannover.de/resource/> .
@prefix so: <https://schema.org/> .
@prefix onyx: <http://www.gsi.dit.upm.es/ontologies/onyx/ns#> .
@prefix dbo: <https://dbpedia.org/ontology/> .

qkg:Quotation584221
 a                      so:Quotation ;
 so:spokenByCharacter   qkg:Person5152 ; # Arshile Gorky
 so:mentions            qkg:Person819 , # Pablo Picasso
                        qkg:Person2707 ; # Paul Cézanne
 qkg:isMisattributed    false ;
 qkg:hasMention         qkg:Mention616243 ; # see below
 onyx:hasEmotionSet     qkg:EmotionSet564951 ; # neutral (score
0.52)
 dbo:year               1945 .

qkg:Mention616243
 a                qkg:Mention ;
 so:text          "I was with C\'{e}zanne for a long time,
                   and now naturally I am with Picasso"@en ;
 so:isPartOf      <https://en.wikiquote.org/wiki/Arshile_Gorky> ;
 so:description   "posthumous"@en ,
                  "Quotes of Arshile Gorky"@en ,
                  "Movements in art since 1945"@en ;
 qkg:hasContext   qkg:Context303407 , # see below
                  qkg:Context303406 . # see below

qkg:Context303407
 a                qkg:Context ;
 qkg:contextText "Quotes of Gorky, from Abstract Expressionist
                  Painting in America, W.C, Seitz, Cambridge
                  Massachusetts, 1983"@en .

qkg:Context303406
 a                qkg:Context ;
 qkg:contextText "p. 31: (in Gorky Memorial Exhibition,
                  Schwabacher pp. 28)"@en .
```

Listing 7.1 Selected triples about a quote by Arshile Gorky

7.3.6 Implementation

We use the MWDumper[17] to process the *Wikiquote* XML dumps and parse the single pages given in the Wikipedia markup using the Bliki engine.[18] For language detection and time expression extraction, we use the langdetect[19] and

[17] https://www.mediawiki.org/wiki/Manual:MWDumper.

[18] https://github.com/axkr/info.bliki.wikipedia_parser.

[19] https://pypi.org/project/langdetect/.

dateparser[20] libraries. The Fast Clustering algorithm is run with an empirically derived cosine similarity threshold of $\tau = 0.8$. The creation of knowledge graph triples and their serialisation is done via the RDFLib library.[21] The Java implementation of the dumper and the Python code for cross-lingual alignment and knowledge graph creation are publicly available on GitHub.[22]

7.4 Statistics, Evaluation, Examples and Web Interface

In this section, we first provide general statistics of *QuoteKG*, evaluate the cross-lingual alignment and present example queries.

7.4.1 Statistics

In total, *QuoteKG* contains 880,878 quotes with 961,535 quote mentions in 55 languages. For 411,912 mentions, context is available. Table 7.2 provides detailed statistics for selected languages. *QuoteKG* covers both high-resource languages such as English (271,541 quote mentions from 19,073 persons) and Italian (146,103 quote mentions from 18,803 persons), as well as low-resource languages such as Welsh (508 quote mentions from 239 persons).

Table 7.2 Statistics of selected languages in *QuoteKG*

Language	Persons	Quotes	Mentions	Mentions with contexts
English	19,073	267,740	271,541	193,848
Italian	18,803	145,235	146,103	48,107
German	3,461	16,012	16,441	4,327
Polish	17,274	119,439	119,880	73,936
Russian	3,954	40,290	41,003	1,331
Hebrew	2,985	48,024	48,330	737
Farsi	3,407	28,101	28,699	8,897
Japanese	488	4,382	4,449	1,603
Croatian	2,707	11,023	12,965	2,045
Welsh	239	461	508	247
All Languages	69,467	880,878	961,535	411,912

[20] https://github.com/scrapinghub/dateparser/.

[21] https://github.com/RDFLib/rdflib.

[22] https://github.com/tkuculo/QuoteKG.

Table 7.3 Evaluation of cross-lingual alignment for eight selected persons in English, German and Italian. TP true positives (mention pairs that were correctly clustered together), TN true negatives (mention pairs that were correctly not clustered together), FP false positives, FN false negatives, P precision, R recall, F_1 F_1 score

Person	TP	TN	FP	FN	P	R	F_1
Alan Turing	10	935	0	1	1	0.91	0.95
Alexander the Great	5	491	0	0	1	1.0	1.0
Edward Snowden	6	697	0	0	1	1.0	1.0
Gustav Mahler	1	44	0	0	1	1.0	1.0
Jean-Claude Juncker	4	776	0	0	1	1.0	1.0
Marie Antoinette	4	347	0	0	1	1.0	1.0
Marie Curie	2	251	0	0	1	1.0	1.0
Tom Clancy	1	2,849	0	0	1	1.0	1.0
Total	33	6,390	0	1	1.0	0.99	0.99

7.4.2 Evaluation of the Cross-lingual Alignment

We evaluate the quality of the cross-lingual alignment of quote mentions by comparing to a ground truth of correctly clustered mentions. Creating such a ground truth is a tedious process due to the large number of possible clusterings and the number of pairwise comparisons.[23] We have selected eight persons with quotes in English, German and Italian and manually clustered their mentions.[24] Ground-truth clusters were then compared to the *QuoteKG* clusters by viewing the clustering process as a series of decisions for each of the pairs of mentions (Schütze et al. 2008). For example, we consider three positive pairs for a quote mentioned in three languages: ($Mention_1$, $Mention_2$), ($Mention_1$, $Mention_3$), ($Mention_2$, $Mention_3$).

Table 7.3 shows the results of this evaluation: Cross-language alignment in *QuoteKG* shows an average precision of 1.0 and an F_1 score of 0.99 for this ground-truth dataset. Following the imbalance of *Wikiquote's* coverage described in Sect. 7.3.1, there is a high number of true negatives, i.e. the majority of quotes are only mentioned once in all *Wikiquote* language versions. In total, there are only two mentions that are not clustered together but should have been. All the other clusters are correct.

Our ground-truth set of manually aligned quote clusters is available on the *QuoteKG* Web site.[3]

[23] When considering a person that has 10 quotes in 5 languages each, there are $\sum_i^{5-1} 10 \times i^2 =$ 1,000 possible pairwise comparisons.

[24] This annotation was performed by the authors who carefully identified which pairs of quotes are translations of each other. The results of this annotation process are available on GitHub: https://github.com/t-kuculo/QuoteKG/tree/main/data/ground_truth.

```
SELECT ?Person (COUNT(?quote) AS ?NumberOfQuotes) WHERE {
  ?quote a so:Quotation ;
    so:spokenByCharacter [
       skos:prefLabel ?Person ] .
} GROUP BY ?Person
ORDER BY DESC(COUNT(?quote))
```

Listing 7.2 SPARQL query for Persons with the most quotes

Table 7.4 The first five results of the query in Listing 7.2

?Person	?NumberOfQuotes
Friedrich Nietzsche	2530
Oscar Wilde	1786
Albert Einstein	1627
Donald Trump	1610
Johann Wolfgang von Goethe	1537

7.4.3 Example Queries

In this section, we present two example queries demonstrating how to use *QuoteKG* as a collection of quotes and as a resource to conduct research on the misattribution of quotes.

7.4.3.1 *QuoteKG* as a Collection of Quotes and Their Originators

Listing 7.2 shows a SPARQL query that returns the five persons with the most quotes in *QuoteKG*. Table 7.4 shows these persons together with the number of quotes. Unsurprisingly, the persons with the most quotes are philosophers and writers, including Friedrich Nietzsche and Oscar Wilde, plus Albert Einstein, known for many (misattributed) quotes (Robinson 2018).

A related example query to use *QuoteKG* for the generation of event quote collections is shown later in Sect. 7.6.1.

7.4.3.2 Verification of Quotes

As also discussed in Chap. 11, misinformation on the Internet has become an increasingly important problem and requires methods that classify the veracity of information (Thorne and Vlachos 2018) and benefit from knowledge graphs such as *ClaimsKG* that provide annotated and erroneous facts (Tchechmedjiev et al. 2019). While *ClaimsKG* provides wrong claims stated by persons extracted from fact-checking sites, *QuoteKG* has quotes labelled as wrongly attributed to persons, thus a different type of misinformation. The query shown in Listing 7.3 returns quotes of Albert Einstein that are marked as misattributed in *QuoteKG* (see Table 7.5),

```
SELECT ?Text (SAMPLE(?contextText) AS ?ContextTexts)
(SAMPLE(?source) AS ?Source) WHERE {
  ?quote so:spokenByCharacter [
    skos:prefLabel "Albert␣Einstein" ] ;
  qkg:isMisattributed true ; qkg:hasMention ?mention .

  ?mention so:text ?Text ;
    qkg:hasContext [
    qkg:contextText ?contextText ; so:source ?source
  ] .
} GROUP BY ?Text
```

Listing 7.3 SPARQL query for Quotes misattributed to Albert Einstein and their contexts

Table 7.5 Two results of the query in Listing 7.3, returning quotes that were misattributed to Albert Einstein. Texts are shortened for brevity

?Text	?ContextTexts	?Source
Everything is energy and that's all there is to it... It can be no other way. This is not philosophy. This is physics	There's no evidence that Einstein ever said this	http://quoteinvesti-gator.com/2012/05/-16/everything-energy/
If the facts don't fit the theory, change the facts	The earliest published attribution of this quote to Einstein found on ..., but no source to Einstein's original writings is given ...	http://books.google.com/books?id=...

together with context information. Such context information can be a valuable resource for explaining misattribution in the case of quotes.

7.4.4 Web Interface

On the *QuoteKG* Web site, we offer a SPARQL endpoint[25] and a demo Search & Demo interface[26] where users can search for specific persons and display their quotes in selected languages. An example of this interface is shown in Fig. 7.7, which displays Portuguese and English quotes of Johann Wolfgang von Goethe.

[25] https://quotekg.l3s.uni-hannover.de/sparql.

[26] https://quotekg.l3s.uni-hannover.de/search.

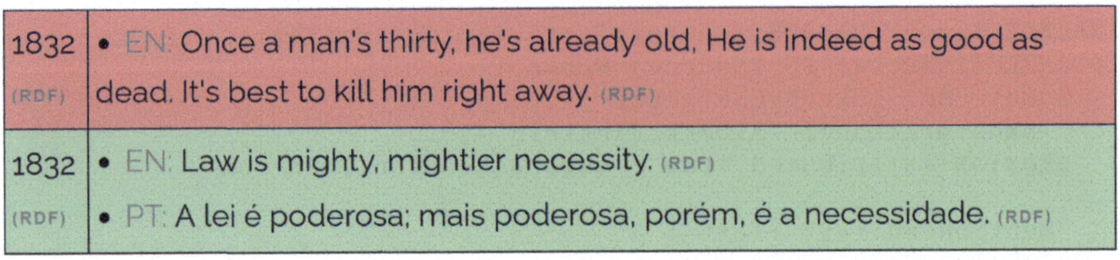

Fig. 7.7 An example of the Search & Demo interface showing two quotes of Johann Wolfgang Goethe, which are available in Portuguese or English. The sentiment of quotes is indicated by colour (red, negative; green, positive)

7.5 Availability

Availability The *QuoteKG* Web site[3] provides access to a description of *QuoteKG* and its schema, to the SPARQL endpoint[25] and to data downloads and will provide a canonical citation to this chapter. *QuoteKG* is licensed under the Creative Commons Attribution-ShareAlike 4.0 International[27] licence. Persistent access to the *Quote-KG* triple files is provided through an upload to the Zenodo repository.[28] The code for the creation of *QuoteKG* is publicly available on GitHub[22] and is licensed under the MIT licence.[29]

Adherence to Standards *QuoteKG* is modelled through the Resource Description Framework. Its schema is an extension of *schema.org*. We provide a machine-readable description of *QuoteKG* using the VoID vocabulary.[30] *QuoteKG* adheres to the linked data principles: resources can be looked up through their URIs, and they are interlinked with *Wikidata* and *DBpedia*.

7.6 Event Analysis Using *QuoteKG*

In this section, we demonstrate the utility of *QuoteKG* in analysing events and its potential to be used together with event-centric knowledge graphs such as *EventKG* (Gottschalk and Demidova 2019) and the *OEKG* (Chap. 6). First, we show how to generate event quote collections from *QuoteKG*. Then, we demonstrate how one can conduct an analysis of public discourse to provide insights into how events shape our society and the importance of exploring them through the lens of quotes. We also illustrate how examining quotes through different time periods can give hints

[27] https://creativecommons.org/licenses/by-sa/4.0/legalcode.
[28] https://zenodo.org/record/4702544.
[29] https://opensource.org/licenses/MIT.
[30] https://www.w3.org/TR/void/.

about the dominant events of the era and that by projecting event-related entities onto a timeline, we can see the levels of interest among the general public in these events. Finally, we investigate the effectiveness of quote data for the exploration of the cultural impact of events through multilingual sentiment analysis.

7.6.1 Generating Event Quote Collections

In order to understand the cultural impact of an event and the ensuing public discourse, researchers can greatly benefit from a comprehensive collection of quotes. These quotes represent the thoughts, opinions and sentiments of individuals, organisations and communities in response to events. However, manually collecting quotes from various sources can be a time-consuming and tedious process, especially when dealing with large-scale events.

In this section, we explore the process of generating quote collections for specific events using *QuoteKG*, highlighting the benefits of using automated methods for event analysis.

Listing 7.4 shows a SPARQL query to generate a quote collection relating to the Syrian civil war,[31] an event in *QuoteKG*. To do so, the query uses the links from quotes to entities represented via `so:mentions`. The extracted quotes in the generated quote collection can be enriched with various attributes, such as sentiment via `onyx:hasEmotionSet` (as in the query, see Sect. 7.2 for more detail), source information and mentions of entities.

Table 7.6 shows four selected quotes contained in the quote event collection about the Syrian civil war collected with our query, together with their negative sentiment and originator. While all negative, the sentiments of quotes clearly differ, with the first one calling the Syrian civil war the "worst humanitarian tragedy" and the least-negative quote among our examples placing hope in upcoming elections. The originators of the quotes cover a wide variety of occupations and nationalities, including politicians, journalists and scholars from Syria, Iran and Europe.

Table 7.7 provides statistics of the event quote collection regarding the Syrian civil war and two more events. To enrich such event quote collections with more quotes, different strategies are applicable, including string searches and filters for relevant persons and event dates.

As seen from our analysis, an event quote collection can provide rich information that can be analysed to gain valuable insights into public opinion and discourse surrounding events. The sentiment expressed in the extracted quotes can be analysed to determine the opinions of different person groups, cultures and nations.

---

[31] https://www.wikidata.org/wiki/Q178810.

```
SELECT ?Text ?Intensity ?Person WHERE {

  ?quotation a so:Quotation .

  ?quotation so:spokenByCharacter [
     skos:prefLabel ?Person
  ] .

  ?quotation so:mentions [
     owl:sameAs dbr:Syrian_civil_war
  ] .

  OPTIONAL {
    ?quotation onyx:hasEmotionSet [
       onyx:hasEmotion [
          onyx:hasEmotionCategory qkg:NegativeEmotion ;
          onyx:hasEmotionIntensity ?Intensity
       ]
    ] .
  }

  ?quotation qkg:hasMention ?mention .
  ?mention so:text ?Text .
}
ORDER BY DESC(?Intensity)
```

Listing 7.4 SPARQL query: Quotes mentioning the Syrian civil war sorted by negative emotion

Table 7.6 Four quotes, their negative sentiment intensity and their authors taken from the Syrian civil war quote collection created with the query in Listing 7.4. Texts are shortened for brevity and translated. We also list the originators' occupations

?Text	?Intensity	?Person
The civil war in Syria is the worst humanitarian tragedy of our generation and one that our government, and the world, is failing to deal with adequately	0.94	Helen Joanne Cox (British politician)
I am against the rampant corruption that gnaws at the system to its very core, against repression and tyranny that obstructs and paralyzes the system, and kills any initiative that might affect the interests of corruption and tyranny. *(translated from Arabic)*	0.89	Mamdouh Hamadeh (Syrian journalist and cartoonist)
He took measures to remove every opponent from his path, first within the party itself, and then outside it. Even if people demonstrated peacefully, the regime's response was expected to be violent	0.70	Nikolaos van Dam (Dutch scholar and author on the Middle East)
The Syrian crisis must be resolved by a vote by Syrians. We are concerned by the civil war and foreign interference. The government must be respected by other countries until the next elections and then it is up to the people to decide	0.57	Hassan Rouhani (seventh president of Iran)

Table 7.7 Statistics of the event quote collections for three selected events

	#Quotes	#Languages	#Persons
Syrian civil war	22	4	21
September 11 attacks	65	8	52
COVID-19 pandemic	33	6	30

7.6.2 *Temporal Analysis of Events Through Quotes*

In this section, we explore the potential of temporal analysis of quotes in uncovering noteworthy information about events. Specifically, we anticipate that this methodology will allow us to identify the particular aspects of an event that capture the public's attention, as well as the manner in which political figures frame said events. Additionally, by analysing the frequency of quotes over time, we may be able to discern patterns in the rise and fall of interest surrounding certain entities, thereby allowing for event detection.

7.6.2.1 Entity Mentions per Year

To test our hypothesis, we conduct a focused examination of quotes across specific time periods. Our analysis centres on the presence of mentioned entities, which we use to determine the salient events of a given period. To do so, we use the property `so:mentions` to identify the most frequently mentioned entities per a given time period.[32]

Our findings indicate that the quotes from the latter half of the nineteenth century, as illustrated in Fig. 7.8a, primarily relate to the American Civil War and the subsequent reconstruction era, as evident by entities such as slavery and the US Constitution. Conversely, Figs. 7.8b and c demonstrate that the year 2016 was defined by the US presidential election (including mentions of Donald Trump, Hillary Clinton and Left-wing politics) and 2020 by the COVID-19 pandemic (COVID-19 pandemic, COVID-19 testing, Anthony Fauci). These figures confirm our assumption that quotes uttered during specific time spans give insights into what people considered relevant during that time and, thus, indicate major events.

7.6.2.2 Timeline of Entity Mentions

We can also utilise the distribution of specific entity mentions over time to identify the peak periods of public interest in these entities and to draw connections to other events. To this end, Fig. 7.9 depicts the six most-mentioned entities between 2000

[32] We skip generic or time-independent entities such as Love, God, Religion, Human, Life, and Jesus.

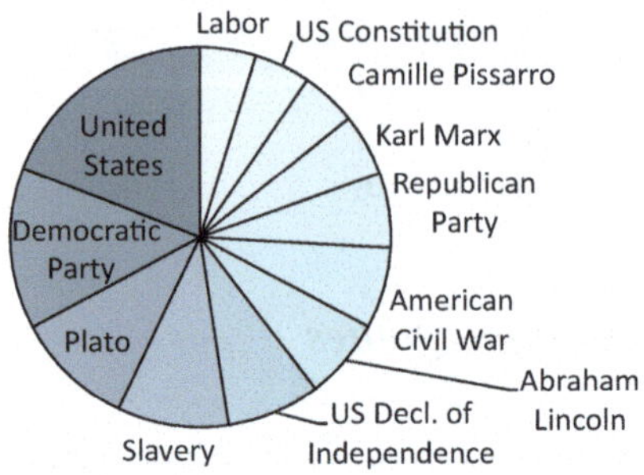

(a) Distribution of entity mentions in quotes from the second half of the 19th century.

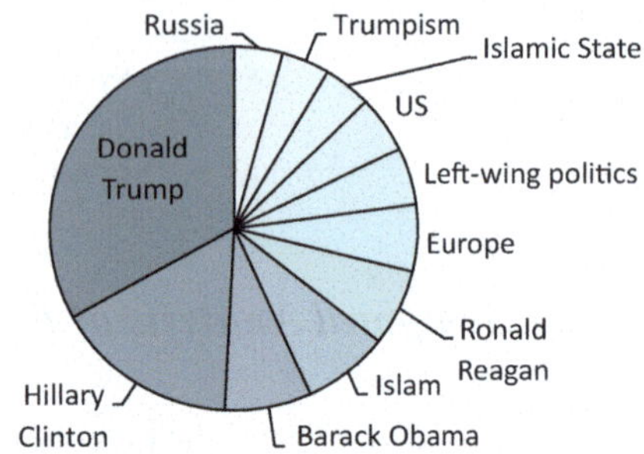

(b) Distribution of entity mentions in quotes from 2016.

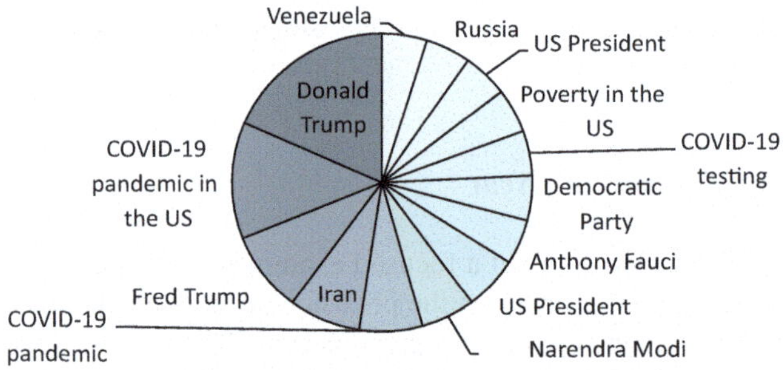

(c) Distribution of entity mentions in quotes from 2020.

Fig. 7.8 Distribution of most-frequent entities mentioned in quotes across different time periods

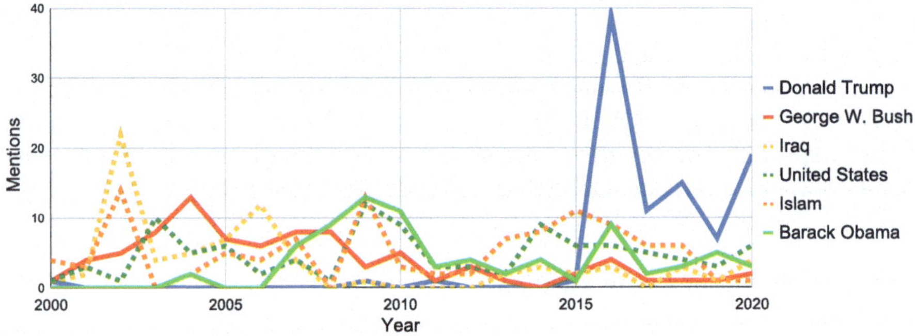

Fig. 7.9 Timeline of entity mentions in the twenty-first century

and 2020 and how often they are mentioned in quotes over time. For example, from that timeline, we can identify the presidencies of three US presidents and topics of relevance such as Iraq during the time of the Iraq War.

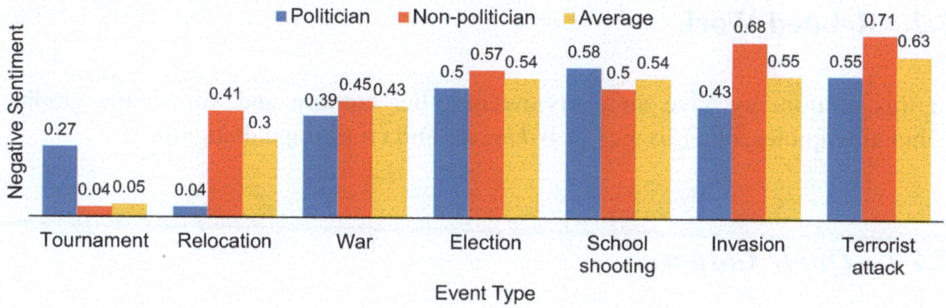

Fig. 7.10 Negative sentiment distribution across event types and per the originator's occupation

Our analyses of entities mentioned in quotes over time suggest that a temporal analysis of quotes is a valuable tool for gaining insight into the public's perceptions and the political framing of historical events.

7.6.3 Exploring the Cultural Impact of Events Through Multilingual Sentiment Analysis

With the ever-growing speed of information propagation, events that have previously affected only select communities now impact peoples of different cultural and linguistic backgrounds. In addition, international and political events, such as conflicts between nations, are often perceived differently by each person according to their beliefs and cultural context.

Multilingual sentiment analysis can offer valuable insights into the cultural impact of events. We assume that the sentiment expressed by different groups of people can shed light on their perspectives and attitudes towards specific events.

Figure 7.10 depicts the negative average sentiment of quotes that mention an event of a specific type (e.g. quotes about tournaments), where we distinguish between quotes said by politicians and by others, i.e. the general public. Quotes mentioning sports tournaments are the least negative, while other event types show a clear negative sentiment, with quotes about terrorist attacks being the most negative ones. We can also observe noticeable differences in sentiment between politicians and the general public for certain event types, such as "Invasion" (0.43 vs 0.68) and "Relocation" (0.04 vs 0.41), with the Israeli disengagement from Gaza, a notable and controversial event, being the only mentioned relocation. These discrepancies may be indicative of conflicting views among different groups and may provide important clues about the cultural impact of these events.

7.7 Related Work

In this section, we give an overview of other corpora and knowledge graphs containing quotes, other usages of *Wikiquote* and cross-lingual alignment.

7.7.1 *Quote Corpora*

Many collections of quotes have been created and maintained, mainly mono-lingual and without semantic annotations. Since the release of its first edition in 1941, The *Oxford Dictionary of Quotations* (Knowles 2009) aims at providing “the wit and wisdom of past and present”, with a focus on the provenance of quotes. The provenance of quotes is also an indispensable criterion in the Book of Fake Quotes (Boller Jr et al. 1989). There are few machine-readable monolingual quote collections[33,34] (Newell et al. 2018; Goel et al. 2018; Vaucher et al. 2021). These corpora are typically monolingual and extracted from the news. Consequently, while they may have a large number of quotes, they lack a mechanism to ensure the societal relevance of quotes, as in *Wikiquote*. As a knowledge graph, *QuoteKG* enables easy access to quotes and rich metadata.

7.7.2 *Quotes and Events in Knowledge Graphs*

While *DBQuote* (Piao and Breslin 2015) allows user annotations of quotes extracted from *Twitter* and *Wikiquote* through an ontology, it only covers two languages (English and Korean) and has not been made available. To the best of our knowledge, *QuoteKG* is the first publicly available knowledge graph of quotes. Consequently, quotes have only been insufficiently covered in the Semantic Web: for example, *Wikidata* (Vrandecic and Krötzsch 2014) contains fewer than 400 instances of the class “Phrase”[35] that are attributed to an author or creator—most of them only consisting of few words (e.g. “cogito ergo sum” and “covfefe”). Event-centric knowledge graphs such as *EventKG* (Gottschalk and Demidova 2019) and the *OEKG* (see Chap. 6) provide an understanding of human history and world-shaking events. They do not include quotes that complement the deeds of public figures. Many applications based on knowledge graphs (e.g. for exploring the lives of persons of public interest (Althoff et al. 2015; Gottschalk and Demidova 2020)) could immediately profit from the inclusion of quotes.

[33] https://www.kaggle.com/akmittal/quotes-dataset.

[34] https://github.com/JamesFT/Database-Quotes-JSON.

[35] https://www.wikidata.org/wiki/Q187931.

7.7.3 *Wikiquote*

Until now, *Wikiquote* has rarely been used as a research corpus, presumably due to the necessary but tedious extraction process that comes with the diverse formatting and template usage both within and across the languages of *Wikiquote*. One example is the work by Buscaldi and Rosso (2008), who manually tagged quotes from the Italian *Wikiquote* as humorous or not and used their annotated corpus for training models for humour recognition. Giammona and Yanes (2019) analysed the spread of ancient quotes in today's Web through *Wikiquote*, and *Wikiquote* was used for training the chatbot Poetwannabe (Chorowski et al. 2018). With *QuoteKG*, we foresee easing access to quotes for a wide range of research questions.

7.7.4 *Cross-lingual Alignment*

Several studies have shown that different languages share similar statistical properties that can be used to learn cross-lingual alignments between two languages, even without relying on any form of bilingual supervision (Chung et al. 2018). While in the past most works and datasets address bilingual alignment (Schamoni et al. 2014; Gottschalk and Demidova 2017; Jing et al. 2019), recently cross-lingual alignment has gathered more attention (Liang et al. 2020). *QuoteKG* focuses on the specific task of cross-lingual alignment of quote mentions.

7.8 Conclusion

In this chapter, we presented *QuoteKG*—a multilingual knowledge graph of quotes—and its application to event analysis. We have presented the *QuoteKG* schema based on `schema.org` as well as a pipeline that extracts quotes from the *Wikiquote* corpus and aligns them across languages. Finally, we demonstrated the creation of event quote collections and conducted analyses on the data, affirming the utility of *QuoteKG* for event analysis. By utilising the described techniques, researchers can efficiently and effectively generate quote collections that capture a range of viewpoints and emotions that emerge in response to events. Subsequent research on the generated data can then be done to provide valuable insights into the cultural impact of events and the resulting public discourse, contributing to a better understanding of the way in which events shape our society. In addition, quotes may be used as a valuable resource to analyse how news spreads across languages (Chap. 10), to support claims (Chap. 11) or to narrativise events (Chap. 12). *QuoteKG* is publicly available and includes nearly one million quotes in 55 languages, said by nearly 69,000 people of public interest.

Acknowledgments The work presented in this chapter has received funding from the European Union's Horizon 2020 research and innovation program under the Marie Skłodowska-Curie grant agreement no. 812997 and under the name CLEOPATRA (Cross-lingual Event-centric Open Analytics Research Academy).

References

Althoff T, Dong XL, Murphy K, Alai S, Dang V, Zhang W (2015) Timemachine: Timeline generation for knowledge-base entities. In: Proceedings of the 21th ACM SIGKDD International Conference on Knowledge Discovery and Data Mining. ACM, pp 19–28

Auer S, Bizer C, Kobilarov G, Lehmann J, Cyganiak R, Ives ZG (2007) DBpedia: A nucleus for a web of open data. In: The Semantic Web, 6th International Semantic Web Conference, 2nd Asian Semantic Web Conference, ISWC 2007. Lecture Notes in Computer Science, vol 4825. Springer, pp 722–735

Barbieri F, Anke LE, Camacho-Collados J (2022) XLM-T: multilingual language models in twitter for sentiment analysis and beyond. In: Proceedings of the Thirteenth Language Resources and Evaluation Conference, LREC 2022. European Language Resources Association, pp 258–266

Boller Jr PF, George Jr OJ, et al. (1989) They never said it: A book of fake quotes, misquotes, and misleading attributions: a book of fake quotes, misquotes, and misleading attributions. Oxford University Press, USA

Buscaldi D, Rosso P (2008) Some experiments in question answering with a disambiguated document collection. In: Evaluating Systems for Multilingual and Multimodal Information Access, 9th Workshop of the Cross-Language Evaluation Forum, CLEF 2008. Lecture Notes in Computer Science, vol 5706. Springer, pp 442–447. https://doi.org/10.1007/978-3-642-04447-2_52

Chorowski J, Lancucki A, Malik S, Pawlikowski M, Rychlikowski P, Zykowski P (2018) A Talker Ensemble: The University of Wroclaw's Entry to the NIPS 2017 Conversational Intelligence Challenge. In: The NIPS'17 Competition: Building Intelligent Systems. Springer, pp 59–77

Chung YA, Lee HY, Glass J (2018) Supervised and unsupervised transfer learning for question answering. In: Proceedings of the North American Chapter of the Association for Computational Linguistics (NAACL). Association for Computational Linguistics

Conneau A, Khandelwal K, Goyal N, Chaudhary V, Wenzek G, Guzmán F, Grave É, Ott M, Zettlemoyer L, Stoyanov V (2020) Unsupervised cross-lingual representation learning at scale. In: Proceedings of the 58th Annual Meeting of the Association for Computational Linguistics, pp 8440–8451

Giammona C, Yanes ES (2019) From Print to Digital Texts, from Digital Texts to Print. Indirect Tradition of Latin Classics on the Web. Storie e Linguaggi Rivista di studi umanistici 1

Goel S, Madhok R, Garg S (2018) Proposing contextually relevant quotes for images. In: Advances in Information Retrieval - 40th European Conference on IR Research, ECIR 2018. Lecture Notes in Computer Science, vol 10772. Springer, pp 591–597

Gottschalk S, Demidova E (2017) Multiwiki: Interlingual text passage alignment in Wikipedia. ACM Trans Web 11(1):6:1–6:30. https://doi.org/10.1145/3004296

Gottschalk S, Demidova E (2019) EventKG - the hub of event knowledge on the web - and biographical timeline generation. Semantic Web 10(6):1039–1070. https://doi.org/10.3233/SW-190355

Gottschalk S, Demidova E (2020) EventKG+BT: Generation of interactive biography timelines from a knowledge graph. In: The Semantic Web: ESWC 2020 Satellite Events - ESWC 2020 Satellite Events. Lecture Notes in Computer Science, vol 12124. Springer, pp 91–97. https://doi.org/10.1007/978-3-030-62327-2_16

Hu J, Ruder S, Siddhant A, Neubig G, Firat O, Johnson M (2020) XTREME: A massively multilingual multi-task benchmark for evaluating cross-lingual generalisation. In: Proceedings of the 37th International Conference on Machine Learning, ICML 2020, PMLR, Proceedings of Machine Learning Research, vol 119, pp 4411–4421

Jing Y, Xiong D, Zhen Y (2019) BiPaR: A bilingual parallel dataset for multilingual and cross-lingual reading comprehension on novels. In: Proceedings of the 2019 Conference on Empirical Methods in Natural Language Processing and the 9th International Joint Conference on Natural Language Processing, EMNLP-IJCNLP 2019. Association for Computational Linguistics, pp 2452–2462

Keyes R (2007) The quote verifier: who said what, where, and when. St. Martin's Griffin

Khurana S (2018) These 4 quotes completely changed the history of the world. https://www.thoughtco.com/quotes-that-changed-history-of-world-2831970

Knowles E (2009) The Oxford dictionary of quotations. Oxford University Press

Krämer A (2021) Ein Satz mit Folgen. https://web.archive.org/web/20220811025404/https://www.tagesschau.de/inland/merkel-wir-schaffen-das-109.html

Kuculo T, Gottschalk S, Demidova E (2022) QuoteKG: A multilingual knowledge graph of quotes. In: The Semantic Web - 19th International Conference, ESWC 2022. Lecture Notes in Computer Science, vol 13261. Springer, pp 353–369

Lehmann J, Isele R, Jakob M, Jentzsch A, Kontokostas D, Mendes PN, Hellmann S, Morsey M, van Kleef P, Auer S, Bizer C (2015) DBpedia - A large-scale, multilingual knowledge base extracted from Wikipedia. Semantic Web 6(2):167–195

Liang Y, Duan N, Gong Y, Wu N, Guo F, Qi W, Gong M, Shou L, Jiang D, Cao G, Fan X, Zhang R, Agrawal R, Cui E, Wei S, Bharti T, Qiao Y, Chen JH, Wu W, Liu S, Yang F, Campos D, Majumder R, Zhou M (2020) XGLUE: A new benchmark dataset for cross-lingual pre-training, understanding and generation. In: Empirical Methods in Natural Language Processing (EMNLP). Association for Computational Linguistics

Mushaben JM (2017) Wir schaffen das! Angela Merkel and the European refugee crisis. German Politics. https://www.tandfonline.com/doi/full/10.1080/09644008.2017.1366988

Newell C, Cowlishaw T, Man D (2018) Quote extraction and analysis for news. In: Proceedings of the Workshop on Data Science, Journalism and Media, KDD

Piao G, Breslin JG (2015) DBQuote: A social web based system for collecting and sharing wisdom quotes. In: Proceedings of the 5th Joint International Semantic Technology Conference, Poster and Demonstrations

Reucher G (2021) Famous quotes: Why are so many fake? https://www.dw.com/en/famous-quotes-why-are-so-many-fake/a-56973281

Robinson A (2018) Did Einstein really say that? Nature. https://www.nature.com/articles/d41586-018-05004-4

Sánchez-Rada JF, Iglesias CA (2016) Onyx: A linked data approach to emotion representation. Inf Process Manag 52(1):99–114

Schamoni S, Hieber F, Sokolov A, Riezler S (2014) Learning translational and knowledge-based similarities from relevance rankings for cross-language retrieval. In: Proceedings of the 52nd Annual Meeting of the Association for Computational Linguistics, ACL 2014. The Association for Computer Linguistics, pp 488–494

Schütze H, Manning CD, Raghavan P (2008) Introduction to information retrieval. Cambridge University Press, Cambridge

Tchechmedjiev A, Fafalios P, Boland K, Gasquet M, Zloch M, Zapilko B, Dietze S, Todorov K (2019) ClaimsKG: A knowledge graph of fact-checked claims. In: The Semantic Web - ISWC 2019 - 18th International Semantic Web Conference. Lecture Notes in Computer Science, vol 11779. Springer, pp 309–324

Thorne J, Vlachos A (2018) Automated fact checking: Task formulations, methods and future directions. In: Proceedings of the 27th International Conference on Computational Linguistics, COLING 2018. Association for Computational Linguistics, pp 3346–3359

Vaucher T, Spitz A, Catasta M, West R (2021) Quotebank: A corpus of quotations from a decade of news. In: WSDM '21, The Fourteenth ACM International Conference on Web Search and Data Mining. ACM, pp 328–336

Vrandecic D, Krötzsch M (2014) Wikidata: a free collaborative knowledgebase. Commun ACM 57(10):78–85. https://doi.org/10.1145/2629489

Chapter 8
Event Recommendation Through Language-Specific User Behaviour in Clickstreams

Sara Abdollahi, Elena Demidova, and Simon Gottschalk

Abstract The relevance and perception of events with global and local impact, such as national elections and terrorist attacks, can vary significantly among different language communities. This chapter discusses recent user access models for event-centric multilingual information, focusing on assisting users, including social scientists and digital humanities researchers, who analyse such events and their impacts. These models aim to facilitate information exploration by emphasising cultural and linguistic differences, a dimension often overlooked by existing entity recommendation methods. Developing recommendation models supporting cross-lingual and cross-cultural analysis of event-related information is particularly challenging due to language barriers and the lack of established datasets. To address these challenges, our prior work involved the creation of the *EventKG+Click* dataset, which contains event-centric user interaction traces extracted from the *EventKG* knowledge graph and Wikipedia clickstream data. Additionally, we introduced *LaSER*—a language-specific event recommendation model that considers the user's linguistic and cultural preferences. To improve recommendations, *LaSER* incorporates language-specific click data from *EventKG+Click*. Furthermore, *LaSER* integrates language-specific embeddings of entities and events, along with their spatio-temporal features, into a learning-to-rank model. This chapter provides an overview of these methods, datasets and evaluation results.

S. Abdollahi (✉) · S. Gottschalk
L3S Research Center, Leibniz Universität Hannover, Hannover, Germany
e-mail: abdollahi@L3S.de; gottschalk@L3S.de

E. Demidova
Data Science & Intelligent Systems Group (DSIS), University of Bonn, Bonn, Germany

Lamarr Institute for Machine Learning and Artificial Intelligence, Bonn, Germany
e-mail: elena.demidova@cs.uni-bonn.de

I. Marenzi et al. (eds.), *Event Analytics across Languages and Communities*,
https://doi.org/10.1007/978-3-031-64451-1_8

8.1 Introduction

Researchers and professionals from various disciplines increasingly recognise the importance of cross-lingual and cross-cultural analytics in the global digital world. This trend is driven by the growing number of events with substantial international impact. Notable examples of such events encompass the COVID-19 outbreak, the European migration crisis and the Brexit referendum. However, language barriers often hinder cross-cultural and cross-lingual research on event-centric information.

Language-specific event and entity recommendations can empower users, including researchers and journalists, to explore and navigate the Web more effectively and investigate topics across various language contexts. Entity recommendation is commonly defined as the task of suggesting entities relevant within a given context, mostly provided as an entity of interest (Ni et al. 2020). Such recommendations can support applications such as Web navigation and exploratory research on user-selected topics. Entity recommendation has been approached from different perspectives, including time-aware entity recommendation (Zhang et al. 2016) and personalised recommendations for social events (Khrouf and Troncy 2013). However, current approaches to event and entity recommendation (Blanco et al. 2013; Ni et al. 2020; Zhang et al. 2016; Bi et al. 2015) typically yield results that do not adequately account for event-specific properties and language context. Moreover, datasets and benchmarks suitable for training and evaluating user interaction methods for cross-lingual information are largely missing. Hence, it is critical to develop approaches that can effectively and efficiently address users' needs in the context of cross-cultural and cross-lingual research. Focusing on the language-specific context, we add novel dimensions to event recommendation.

In Table 8.1, we present examples of events particularly relevant in the context of the coronavirus pandemic from the perspective of the German-, Italian- and Spanish-speaking Wikipedia audiences. These examples are derived based on the number of clicks in the Wikipedia Clickstream dataset[1] that provides the click-through rates for Wikipedia articles in their respective Wikipedia language edition. As we can observe, according to Wikipedia Clickstream, German Wikipedia users are predominantly interested in coronavirus outbreaks in Germany and the economic recession triggered by the pandemic in German-speaking countries. Italian users are mainly interested in the pandemic in Italy and the SARS outbreak, a similar event in 2002, followed by the COVID-19 pandemic in the USA. In contrast, Spanish Wikipedia reflects user interests spanning several Spanish-speaking countries, including Argentina, the USA and Colombia. These observations illustrate how event relevance can vary depending on the language-specific and cultural user contexts.

This chapter provides an overview of our recent contributions to language-specific event recommendations. In particular, it encompasses the development

[1] https://meta.wikimedia.org/wiki/Research:Wikipedia_clickstream

Table 8.1 Events with the highest number of clicks in the German, Italian and Spanish Wikipedia starting from the article *Coronavirus pandemic* in April 2021 (*N*: The number of user clicks in the respective Wikipedia language edition)

German	*N*	Italian	*N*	Spanish	*N*
COVID-19 pandemic in Germany	3,775	COVID-19 pandemic in Italy	2,890	COVID-19 pandemic in Argentina	1,452
COVID-19 recession	1,852	2002–2004 SARS outbreak	780	COVID-19 pandemic in the United States	1,286
COVID-19 pandemic in Austria	1,072	COVID-19 pandemic in the United States	489	COVID-19 pandemic in Colombia	1,105

of the *EventKG+Click* dataset, initially introduced in Abdollahi et al. (2020), as well as our approach to the task of language-specific event recommendation (*LaSER*) as first described in Abdollahi et al. (2023). With *EventKG+Click* and *LaSER*, we add two crucial dimensions to entity recommendation: (i) the ability to recommend events of societal significance, such as political elections and sports events and (ii) the incorporation of language-specific context for these events. These dimensions are essential in various application scenarios, including event recommendation within information retrieval, event analytics for understanding cultural perspectives (Gottschalk et al. 2018) and exploring the diverse perceptions of events across different cultures (Liu et al. 2005). For example, Table 8.2 shows the highest ranked events per language in EventKG+Click. *LaSER* is trained on publicly available data, moving away from dependency on proprietary click or search logs, as frequently utilised in the literature, e.g. Ni et al. (2020) and Huang et al. (2018).

LaSER is based on language-specific latent representations (embeddings) of entities and events in a language-specific knowledge graph representing the relevance of entity and event relations within different language contexts. We combine these latent representations with spatio-temporal event features and utilise them for training a learning-to-rank (LTR) model. Given a language of interest and a query entity, this model generates a ranked list of relevant events. We train *LaSER* on our publicly available *EventKG+Click* dataset that we also integrated into the OEKG as described in Chap. 6. *EventKG+Click* aims to provide a reference source

Table 8.2 Events with the highest language-specific relevance per language in *EventKG+Click*

Rank	English	German	Russian
1	Southeast Asian Games	2016 Berlin truck attack	2009 Russian Premier League
2	2017 Southeast Asian Games	German student movement	1993 Russian Top League
3	2014 United States Senate elections	2006 Austrian legislative election	2012–13 Russian Premier League

for training and evaluating models for event-centric cross-lingual user interaction. Furthermore, we supplement *EventKG+Click* with high-quality user relevance judgements obtained through a user study.

We evaluated the effectiveness of *LaSER* in two distinct settings. Firstly, we evaluated *LaSER*'s capacity to predict language-specific clicks between entities and events in the Wikipedia Clickstream dataset. The results demonstrated that our model outperforms link-based, embedding-based and graph attention network-based ranking baselines by over 8 (nDCG@10) and 17 (MAP@10) percentage points on average. Secondly, we conducted a user study to evaluate the relevance of recommended events and analyse various relevance criteria. The outcomes of this study confirmed that *LaSER* outperforms the baselines by up to 33 percentage points in MAP@5 concerning the language-specific relevance.

8.2 Related Work

The *LaSER* approach summarised in this chapter aims at recommending events while considering their language-specific relevance. This section describes related research areas: entity and event recommendation, learning to rank and graph embeddings.

8.2.1 Entity and Event Recommendation

While user-item and entity recommendation tasks have been extensively studied in the literature, event recommendation has until now been mainly focused on social media events (Gao et al. 2016; Qiao et al. 2014).

8.2.1.1 Entity Recommendation

Entity recommendation is recommending a ranked list of entities to the user query. Blanco et al. (2013) presented Spark, an entity recommendation system that, by using and combining several signals from various data sources, ranks the entities related to the user query. Ni et al. (2020) proposed a framework for recommending related Wikipedia entities using an architecture of multiple layered graphs, candidate generation via *Doc2Vec* embeddings and ranking with an *LTR* model. Other approaches focus on specific recommendation aspects: Zhang et al. (2016) proposed a time-aware entity recommendation (*TER*), which allows users to restrict their interests in entities to a customised time range. Tran et al. (2017) extended *TER* by incorporating topic and time and proposed contextual relatedness among entities using embedding techniques. User interest and preference have also been studied by Bi et al. (2015), who proposed the "probabilistic Three-way Entity

Model" (*TEM*) that provides personalised recommendations of related entities using user interactions from personal click logs. Huang et al. (2018) studied serendipity to engage users' interest while recommending entities.

Existing entity recommendation methods focus on recommendations regardless of language preferences and often rely on proprietary search log data, making them challenging to reproduce. In contrast, our proposed *LaSER* recommends relevant language-specific events and relies on open data, addressing significant limitations of existing methods.

8.2.1.2 Event Recommendation

Events play an important role in a range of real-world applications, including news search (Rudnik et al. 2019), news linking (Setty and Hose 2018) and event-centric user interfaces (Gottschalk and Demidova 2018). However, event recommendation has not been extensively studied and primarily focuses on social media events. Existing event recommendation approaches (Gao et al. 2016; Qiao et al. 2014) focus on event-based social networks (ESRN) such as Meetup, where the goal is to recommend social events such as parties, concerts and conferences to the users.

Unlike the approaches mentioned above, we focus on events of societal importance, such as the coronavirus pandemic and the Second World War. With the proposed *LaSER* approach, we leverage structured information from knowledge graphs and consider information needs and the context of language communities.

8.2.2 Learning to Rank

The ranking is essential for many recommendation algorithms, typically following the candidate generation step. Given a set of objects, a ranking model calculates the score of each object and sorts them accordingly. The scores may represent the degrees of relevance, preferences or importance, depending on applications (Liu et al. 2009). LambdaMART (Burges 2010) is a recent *LTR* model that uses a boosted tree model. LambdaMART demonstrated superior performance when click-based data were used as features (Wu et al. 2018) and has been applied in many application domains, including recommendations (Palumbo et al. 2017), e-commerce click and search (Guo et al. 2020b). Our *LaSER* approach relies on LambdaMART as a basis for learning to rank.

8.2.3 Embedding Methods

Graph embedding techniques have recently been adopted for recommendation tasks (AlGhamdi et al. 2021) and aim to represent graph nodes by low-dimensional vectors, created using random-walk-based (Perozzi et al. 2014; Grover and Leskovec 2016), deep-learning-based (Cao et al. 2016) and factorisation-based methods (Ou et al. 2016; Goyal and Ferrara 2018).

Knowledge graph embeddings specifically target the embedding of entities and relations in a knowledge graph. Translational distance models such as *TransE* (Bordes et al. 2013) and its extensions exploit distance-based scoring functions. They measure the plausibility of a fact as the distance between the two entities, usually after a translation carried out by the relation (Wang et al. 2017). Other knowledge graph embedding methods employ additional information such as entity types (Xie et al. 2016), relation paths (Toutanova et al. 2016), textual information (Xie et al. 2016) and hybrid information (e.g. *Wikipedia2Vec* Yamada et al. 2020) in the embedding process. In Sect. 8.6.1, we discuss the impact of different knowledge graph embedding methods on *LaSER* performance and the benefits of using language-specific embeddings.

8.3 Problem Statement

This section defines the notions of a language-specific knowledge graph, entities and events and the task of language-specific event recommendation.

To facilitate recommendation, we introduce a language-specific knowledge graph, which models entities, events and relations in a language context.[2]

Definition 1 A **language-specific knowledge graph** is a directed graph $\mathcal{G} = (\mathbb{V}, \mathbb{E}, \mathbb{L})$ whose vertices $\mathbb{V}$ represent a set of real-world entities (e.g. persons, places and events), connected via edges $\mathbb{E} \subset \mathbb{V} \times \mathbb{V}$. The set of languages is denoted as $\mathbb{L}$.

Following this definition, the language-specific knowledge graph contains information regarding a set of languages ($\mathbb{L}$) to serve as background knowledge for language-specific event recommendation. Even when only interested in a specific language $l \in \mathbb{L}$, information available in other languages is still relevant and can also help estimate the popularity of an entity in relation to other languages.

In the context of language-specific recommendations, relevant spatio-temporal features are locations and dates associated with entities.

Definition 2 An entity $v \in \mathbb{V}$ can be assigned a start and end time $[t_s(v), t_e(v)]$ as well as a set of coordinate pairs $\mathbb{C}(v)$, where each coordinate pair $c \in \mathbb{C}(v)$ consists of latitude and longitude: $c = (lat, lon), lat \in \mathbb{R}, lon \in \mathbb{R}$.

For example, the *Summer Olympics 2012* happened from 27 July to 12 August 2012 and are assigned multiple coordinate pairs reflecting different sports venues in

[2] Note that in this work, we follow a language-specific view, i.e. we do not further distinguish between different sub-communities speaking the same language (e.g. the different English-speaking sub-communities).

London. The entity representing *Winston Churchill* is assigned his birth and death dates (30 November 1874 to 24 January 1965) and a set of coordinate pairs referring to essential places in his life (e.g. of Blenheim Palace, his birthplace).

In the context of the language-specific knowledge graph, events are a subset of entities. Whereas many definitions of an event exist in the literature, in this work, we follow an event definition by Allan et al. (1998) proposed in the context of event detection and tracking within news stories:

Definition 3 An event $u \in \mathbb{U} \subset \mathbb{V}$ is something that happened at a particular time and place.

Examples of events are the *Summer Olympics 2012*, the *fire at Notre Dame* in 2020 and the *coronavirus pandemic in Germany* starting in 2020. The end date may be unknown for ongoing events like the *coronavirus pandemic*.

Having introduced the entities, events and their relations, we can now define the task of language-specific event recommendation.

Definition 4 Given a query entity $v \in \mathbb{V}$, a language $l \in \mathbb{L}$ and the language-specific knowledge graph $\mathcal{G} = (\mathbb{V}, \mathbb{E}, \mathbb{L})$, the task of **language-specific event recommendation** is to create a sequence $S_{v,l} = (u_1, \ldots u_n)$ of events $(u_i \in \mathbb{U}, i \in 1, \ldots, n)$. The events in $S_{v,l}$ are sorted in descending order regarding their relevance to the query entity v for the audience speaking the language l.

For example, consider the recommendation example in Table 8.1 created from the click counts on Wikipedia articles in specific Wikipedia language editions in April 2021. For the query entity $v =$ *Coronaviruspandemic* and $l =$ *German*, this method returns a list $S_{v,l}$ of recommended events (*COVID-19 pandemic in Germany, COVID-19 recession, COVID-19 pandemic in Austria*). Language-specific event recommendation generates a ranked list of events. The query entity may be any node in the language-specific knowledge graph.

8.4 The *LaSER* Approach

In this section, we summarise *LaSER*, a method for language-specific event recommendation first introduced in Abdollahi et al. (2023). Figure 8.1 provides an overview of the *LaSER* components. *LaSER* consists of a training and a query phase. These phases rely on background knowledge, namely, the language-specific knowledge graph and language-specific click data (*EventKG+Click*; Abdollahi et al. (2020)). In the pre-processing training phase, we create language-specific embeddings based on the language-specific knowledge graph. In addition, we train a learning-to-rank model that learns from language-specific click data. This model uses feature values extracted from the language-specific knowledge graph, i.e. event characteristics and the relationships between events and entities. In the query phase, given an input query entity $v \in \mathbb{V}$ and a language $l \in \mathbb{L}$, we use the embeddings and the trained learning-to-rank model to generate a ranked list of events. In this section,

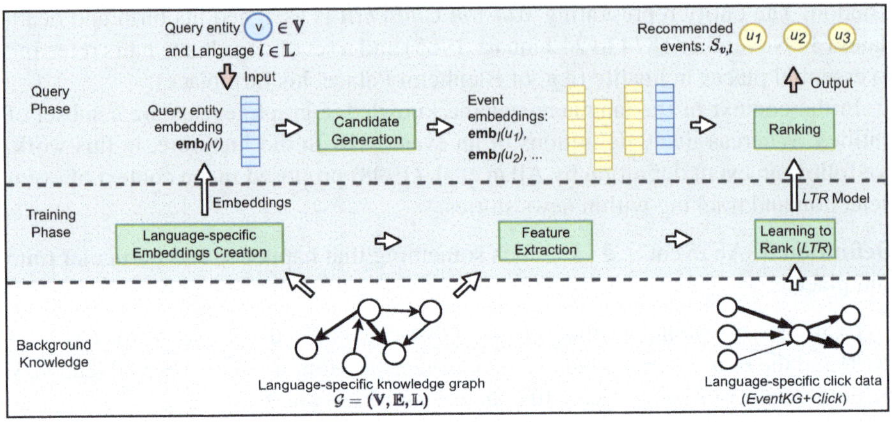

Fig. 8.1 The *LaSER* overview includes three parts. (i) The background knowledge includes the language-specific knowledge graph and the language-specific click data. (ii) In the training pre-processing phase, the language-specific embeddings and the *LTR* event ranking model are trained based on this background knowledge. (iii) In the query phase, given a query entity v (e.g. *coronavirus pandemic*) and a language l (e.g. German) as an input, the embeddings and the *LTR* ranking model are utilised to generate a language-specific ranked list of events $S_{v,l}$ (e.g. (*COVID-19 pandemic in Germany*, *COVID-19 recession*, *COVID-19 pandemic in Austria*))

we describe the background knowledge, training and query phases of *LaSER* in more detail.

8.4.1 Background Knowledge

The *LaSER* approach relies on background knowledge, including the language-specific knowledge graph and language-specific click data.

8.4.1.1 Language-Specific Knowledge Graph

Following Definition 1, the language-specific knowledge graph $\mathcal{G} = (\mathbb{V}, \mathbb{E}, \mathbb{L})$ represents entities, their spatial and temporal characteristics and relationships in the context of a specific language $l \in \mathbb{L}$.

8.4.1.2 Language-Specific Click Data: *EventKG+Click*

The language-specific click data provides training labels for the ranking model. We utilise our *EventKG+Click* dataset (Abdollahi et al. 2020) extracted from the Wikipedia Clickstream, representing real user interactions with Wikipedia articles corresponding to the entities in the language-specific knowledge graph. From such user interactions, we infer the language-specific interaction-based relevance scores for an entity $v \in \mathbb{V}$ and an event $u \in \mathbb{U}$: $ri(v, u, l) \in [0, 1]$. Wikipedia language versions have differing numbers of active users, edits and articles, resulting in an imbalance in the number of clicks in their clickstreams. Therefore, to observe language-specific behaviour, the effects that originate from the popularity of specific language versions need to be balanced. To obtain balanced clicks cb, click counts c are multiplied by the relative number of clicks in the particular language version.

$$cb(v_s, v_t, l) = c(v_s, v_t, l) \times \frac{\sum_{l' \in \mathbb{L}} \sum_{v'_s \in \mathbb{V}} \sum_{v'_t \in \mathbb{V}} c(v'_s, v'_t, l')}{\sum_{v'_s \in \mathbb{V}} \sum_{v'_t \in \mathbb{V}} c(v'_s, v'_t, l)} \tag{8.1}$$

The language-specific relation r is computed as the fraction of clicks in the given language compared to all languages:

$$r(v_s, v_t, l) = \frac{cb(v_s, v_t, l)}{\sum_{l' \in \mathbb{L}} cb(v_s, v_t, l')} \in [0, 1] \tag{8.2}$$

Note that this metric rules out the effects resulting from the relevance of the entity: Events highly related to an entity v can obtain relevance scores close to 1 independent of v's click count.

8.4.2 Training Phase

The training phase aims to create language-specific embeddings and train an event ranking model. The training phase is conducted as a pre-processing and does not impact the query efficiency. This phase consists of the following three steps:

1. Language-specific embeddings creation: From the language-specific knowledge graph $\mathcal{G} = (\mathbb{V}, \mathbb{E}, \mathbb{L})$, we create language-specific embeddings of entities and events such that for each entity $v \in \mathbb{V}$ and each language $l \in \mathbb{L}$, there is an embedding vector $\mathbf{emb}_l(v)$.
2. Feature extraction: For a pair of an entity and an event, we extract feature values representing different characteristics of the event and the pair. Example features are the event popularity, the spatial distance between the entity and the event and their embedding similarity.
3. Learning to rank: We incorporate the features to train an *LTR* model, which ranks events regarding their relevance to the query entity.

In the following, we describe these steps in more detail.

8.4.2.1 Language-Specific Embeddings Creation

To efficiently leverage entities' information and the structure of the language-specific knowledge graph, we provide the overview of a language-specific embedding technique that learns entity embeddings $\mathbf{emb}_l$ representing their relations in a language l. This technique maps the entities to low-dimensional vectors, which are similar when two entities appear close to each other in the language-specific context of l.

To create language-specific embeddings, we utilise *DeepWalk* (Perozzi et al. 2014) and follow a uniform random sampling approach. After creating a set of random walks following the *DeepWalk* approach, we train a *Word2Vec* model. The resulting language-specific embeddings are utilised for (i) creating a candidate set of events relevant to the query entity and (ii) measuring the language-specific relevance between the query entity and the event.

8.4.2.2 Feature Extraction

To model event relevance to the query entity v in the context of language l, we extract ten feature values from the event u and the entity v. This set of features includes four groups, covering different entity aspects: spatial, temporal, link-based and embedding-based features.

- **Spatial features:** Language distance, Pair distance
- **Temporal features:** Interval overlap, Begin time distance
- **Link-based features:** Number of incoming links, Number of outgoing links, Number of shared incoming links, Number of shared outgoing links, Milne-Witten relatedness
- **Embedding-based features:** Embedding similarity

Detailed descriptions of these features are provided in Abdollahi et al. (2023).

8.4.2.3 Learning-to-Rank

To rank the events relevant to the query entity, we train a learning-to-rank model that takes feature values as input and is trained to predict the ranking inferred from the language-specific click data. In the context of the *LTR* model, the problem of language-specific event recommendation is defined as follows: given a training set of language-specific relevance values between entities and events as well as their features, learn a scoring function that approximates the language-specific interaction-based relevance $ri(v, u, l)$ for the query entity v and an event u in a language l.

We train a tree ensemble model to learn an optimal ranking of the language-specific relevance scores using LambdaMART (Burges 2010). LambdaMART is an *LTR* algorithm that uses gradient-boosted decision trees with a cross-entropy cost function. In the literature, LambdaMART has been shown to outperform several neural ranking models in information retrieval tasks (Guo et al. 2020a). Using LambdaMART, we perform a list-wise ranking where the normalised discounted cumulative gain (nDCG) is maximised.

8.4.3 Query Phase

In the query phase, *LaSER* takes the query entity $v \in \mathbb{V}$ and a language $l \in \mathbb{L}$ given by the user as input and recommends a language-specific ranking of events $S_{v,l}$ as an output.[3]

The query phase consists of the following two steps:

1. Candidate generation: A set of candidate events is generated based on the language-specific embeddings.
2. Ranking: The candidate events are ranked by the previously trained *LTR* model.

8.4.3.1 Candidate Generation

Due to the unfeasibility of ranking all events' relevance, a candidate generation strategy is required. In line with the concept of candidate generation in Ni et al. (2020), we select $k = 200$ (as described in Sect. 8.6.1) events most similar to the query entity v based on the embedding similarity.

8.4.3.2 Ranking

Finally, for each candidate event u, we compute its feature values as well as the feature values between the query entity v and u based on the language-specific knowledge graph. Given the input set of all candidate events and their feature values, we employ the *LTR* model trained in the training phase to estimate the language-specific relevance scores. We utilise the resulting scores to sort the candidate events according to their relevance and create the set of recommendations $S_{v,l}$.

[3] We assume that users can select a query entity from the language-specific knowledge graph, e.g. via its label.

8.5 Evaluation Setup

This section describes the ground truth for evaluating event recommendations by *LaSER*, presents embedding and recommendation baselines, describes the evaluation metrics and provides the implementation details.

8.5.1 Ground-Truth Creation

To train *LaSER* and evaluate the language-specific recommendations, we automatically derived a ground truth of language-specific event recommendations from *EventKG+Click*. For each considered language l, this ground-truth $\mathbb{G}_l$ contains query entities together with a ranked list of events and is composed as follows:

$$\begin{aligned}\mathbb{G}_l = &\{(v, (u_1, \ldots, u_n, u_1^-, \ldots, u_n^-)) \\ &| ri(v, u_i, l) \geq ri(v, u_j, l) \\ &\forall 1 \leq i < j \leq n\},\end{aligned} \tag{8.3}$$

where u_i^- denotes negative examples, i.e. randomly chosen events unrelated to the query entity: $ri(v, u_i^-, l) = 0$. In other words, we select all entities in *EventKG+-Click* for which events are provided and rank these events according to their ri score. Each ranked event list is complemented with randomly selected negative examples of the same number as the positive event examples.

The resulting ground-truth dataset is available online.[4] The statistics of this dataset are presented in Table 8.3.

8.5.2 Embedding Methods

LaSER relies on node embeddings for candidate generation and adopts them as a ranking feature. We compare the following embedding methods in our evaluation:

- *DeepWalk* node embeddings (Perozzi et al. 2014) are learnt by applying neural language models to random walks treated as sentences.

Table 8.3 Statistics of the language-specific click data obtained from *EventKG+Click*

	German	French	Russian
Source Entities	117,281	104,331	97,212
Events	40,223	46,557	33,712
Relevance Pairs	304,564	271,243	254,910

[4] https://zenodo.org/record/5735580

- *Node2Vec* node embeddings (Grover and Leskovec 2016) are learnt through biased random walks that consider edge weights for flexible exploration of network neighbourhoods.
- *Wikipedia2Vec* word and entity embeddings (Yamada et al. 2020) are learnt using the skip-gram model on Wikipedia's link graph, texts and link context terms.
- *TransE* node embeddings (Bordes et al. 2013) are inferred by interpreting relationships as translations operating on low-dimensional entity embeddings.

8.5.3 Recommendation Baselines

To compare the *LaSER* approach to the state-of-the-art recommendation baselines, we need to ensure that the baselines (i) represent the state of the art in the entity or event recommendation, (ii) can be applied to the novel task of language-specific event recommendation considered in this work and (iii) are reproducible, i.e. do not depend on any proprietary data.

Therefore, following the evaluation procedure in Tran et al. (2017), we evaluate *LaSER* against four recommendation baselines, Milne-Witten (Witten and Milne 2008), *DeepWalk* (Perozzi et al. 2014) and *Node2Vec* (Grover and Leskovec 2016), as well as SuperGAT (Kim and Oh 2021). The relevance score provided by each baseline between an entity $v \in \mathbb{V}$ and an event $u \in \mathbb{U}$ in a language l is used for event ranking in language-specific event recommendation.

8.6 Evaluation

This section describes five evaluation steps: assessing the impact of language-specific embedding on the candidate generation step of *LaSER*, evaluation of the recommendations, feature analysis, a user study and an anecdotal evaluation.

8.6.1 Candidate Generation Evaluation

As illustrated in Fig. 8.1, the *LaSER* query phase consists of two main steps: candidate generation and ranking. In this experiment, we evaluate *LaSER*'s performance on the candidate generation task based on the ground truth described in Sect. 8.5.1, limited to those cases where a query entity has more than ten clicked target events.

As described in Sect. 8.4.3.1, given a query entity v, the candidate generation step retrieves a set of candidate events regarding their embedding similarity towards v. We compare the performance of different embedding methods to demonstrate the effectiveness of *LaSER*'s language-specific embeddings for candidate generation. We utilise the embedding techniques provided in Sect. 8.5.2.

Table 8.4 Candidate Recall achieved by the *LaSER* utilising different embedding methods

Model	Candidate Recall			
	German	French	Russian	Average
Deepwalk	**0.408**	**0.312**	**0.373**	**0.364**
Node2Vec	0.348	0.276	0.371	0.332
TransE	0.009	0.007	0.008	0.008
Wikipedia2Vec	0.017	0.017	0.018	0.017

For each query entity in the ground truth and each embedding technique, we retrieve the 200 most similar events as candidate events (see Sect. 8.4.3), i.e. less than 500 candidate entities as in Ni et al. (2020) because there are fewer events than entities in the knowledge graph. Then, we compute the candidate recall per embedding technique, i.e. the fraction of events in the ground truth contained in the candidate events. The results are shown in Table 8.4. The *DeepWalk* and *Node2Vec* embeddings outperform the other two embeddings, with the non-language-specific *TransE* embedding performing worst. This result demonstrates the benefit of creating language-specific random-walk-based embeddings for the language-specific event recommendation.

8.6.2 Recommendation Evaluation

In this experiment, we evaluate *LaSER*'s performance on the recommendation task. Given a query entity and a set of candidate events, this task aims to rank the candidate events according to their relevance to the query entity for the audience speaking the language of interest, i.e. the language-specific relevance.

In this experiment, *LaSER* is trained via a 5-fold cross-validation on each language separately. The folds are created based on the set of query entities: in each run, we use 80% of the query entities and their events in the ground truth for training the *LTR* model, the remainder for testing. The results are averaged over the 5 runs.

To assess recommendation quality, we employ nDCG@10 (normalised discounted cumulative gain at 10) and MAP@10 (mean average precision at 10). nDCG@10 (Järvelin and Kekäläinen 2002) compares the first 10 events in the ranking against the ideal ground-truth ranking, rewarding relevant events in higher positions more. The optimal nDCG@10 score is 1.0. MAP@10 calculates the average precision scores (AP@10) for each query entity in the ground truth, where AP@10 is the sum of precision scores (precision@k for $k = 1 \ldots 10$) divided by the total number of relevant events in the top 10 ranked results.

Table 8.5 reports the nDCG@10 and MAP@10 scores of the recommendation evaluation for the four recommendation baselines and *LaSER* in three languages. As we can observe, in all three languages, *LaSER* clearly outperforms the baselines. On average, across languages, with an nDCG@10 of 0.957 and MAP@10 of

Table 8.5 nDCG@10 and MAP@10 scores achieved by the *LaSER* approach and the recommendation baselines in three languages in the ranking study

	nDCG@10 Score				MAP@10 Score			
	German	French	Russian	Avg.	German	French	Russian	Avg.
Milne-Witten	0.893	0.897	0.890	0.893	0.848	0.864	0.838	0.850
Node2Vec	0.860	0.841	0.885	0.862	0.729	0.679	0.803	0.737
DeepWalk	0.899	0.858	0.901	0.886	0.731	0.850	0.848	0.810
SuperGAT	0.853	0.884	0.879	0.872	0.824	0.806	0.780	0.803
LaSER	**0.957**	**0.958**	**0.956**	**0.957**	**0.969**	**0.970**	**0.971**	**0.970**

Table 8.6 Feature analysis: The results of *LaSER* when removing feature groups. We report the nDCG@10 scores in three languages

Model	German	French	Russian
LaSER	**0.957**	**0.95**	**0.956**
- Spatial features	0.956	0.952	0.955
- Temporal features	**0.956**	0.957	0.956
- Link-based features	0.911	0.946	0.909
- Embedding-based features	0.950	0.957	0.951

0.97, *LaSER* outperforms the baselines by more than 8 and 17 percentage points, respectively. The *LaSER* performance is similar across languages.

8.6.3 Feature Analysis

We perform a feature analysis to assess the effectiveness of specific feature groups in *LaSER*. To this extent, we remove one feature group at a time and measure the resulting performance regarding nDCG@10. The results are presented in Table 8.6. As we can observe, each feature group contributes towards the *LaSER* overall performance. The link-based features provide the highest contribution among the four feature groups, while the temporal features have the lowest impact. We observe similar effects of feature groups across all languages.

A relatively low contribution of the spatial and temporal features can be explained through the non-availability of these features for a large proportion of entities. Furthermore, whereas language-specific embeddings provide a substantial contribution in the candidate generation step, as discussed in Sect. 8.6.1, they have only a limited impact on the follow-up ranking step. An average embedding-based similarity in this step is 0.65 with a relatively low standard deviation of $\sigma = 0.18$. Thus, re-ranking candidates based on the embedding-based similarity is only possible to a limited extent. Overall, incorporating all the provided feature groups leads to the best performance of the approach.

8.6.4 User Study

The user study aims to assess the recommendation quality from the user perspective. The evaluation of recommendation methods is often accomplished through pooling, as existing datasets like Wikipedia Clickstream only cover a portion of potentially relevant events to a query entity. To conduct the user study, we selected ten popular query entities of various types: events, places, persons, art and religion. In the study, 17 post-graduate researchers in computer science and digital humanities participated, collectively annotating 935 triples. Each participant annotated at least 9 triples, with an average of 55 triples per participant.

The study breaks down the judgement of event relevance into three criteria to gain more detailed insights: relevance to the (i) topic, (ii) language community audience and (iii) the general audience. If an event is relevant, at least regarding (i) and (ii), we consider it to be of language-specific relevance. Further details of the user study are reported in Abdollahi et al. (2023).

Table 8.7 illustrates that *LaSER* outperforms the recommendation baselines in most relevance criteria and languages, notably achieving the highest MAP@5 on relevance to the language audience in all languages. In contrast, all approaches achieve high MAP@5 scores for topic relevance, where the two baselines slightly outperform *LaSER*.

Table 8.7 User study results: MAP@5 of *LaSER* and two recommendation baselines in three languages regarding three relevance criteria judged in the user study and the overall language-specific relevance

	Relevance (MAP@5)			
	Topic	General Audience	Language Audience	Language-specific Relevance
German				
Milne-Witten	**1.00**	0.87	0.81	**0.81**
DeepWalk	0.98	0.77	0.70	0.70
LaSER	0.93	**0.89**	**0.91**	**0.81**
French				
Milne-Witten	**1.00**	0.88	0.68	0.68
DeepWalk	**1.00**	0.88	0.78	0.77
LaSER	0.95	**0.94**	**0.95**	**0.90**
Russian				
Milne-Witten	**1.00**	0.90	0.59	0.59
DeepWalk	**1.00**	**0.95**	0.51	0.51
LaSER	0.98	0.90	**0.84**	**0.84**

Table 8.8 Events recommended for the query entity *Film Festival* in three languages

German	French	Russian
46th Venice International Film Festival	34th César Awards	All-Union Film Festival
International Short Film Festival Oberhausen	6th César Awards	Окно в Европу (кинофестиваль) / *Window to Europe (film festival)*
KALIBER35 Munich International Short Film Festival	21st Lumières Awards	Moscow International Film Festival
Filmfest Hamburg	17th César Awards	Short Film
Filmfest München	Brest European Short Film Festival	Kinotavr

8.7 Anecdotal Result

In our final evaluation step, we analyse selected event recommendations of *LaSER* for one query entity annotated during the user study. In this section, we present an anecdotal result to highlight the strengths of the *LaSER* approach.

As our example query entity, we select *Film Festival*. The top 5 events recommended by *LaSER* for German, French and Russian are shown in Table 8.8. The recommended events clearly show a language-specific focus, i.e. important film festivals that happen in cities where the respective languages are spoken: for example, *LaSER* recommends *Filmfest München* for the German audience, several César Awards for the French audience and Окно в Европу, a Russian film festival happening in the Russian city *Wyborg*, for the Russian audience.

In general, this anecdotal example further illustrates that *LaSER* can recommend relevant events that differ across languages, reflecting language-specific relevance.

8.8 Conclusion

In this chapter, we provided an overview of *LaSER*, a method for language-specific event recommendation. *LaSER* leverages language-specific entity embeddings and a learning-to-rank model to generate a list of events relevant to a given query entity within a language-specific context. As language-specific click data, we adopt the *EventKG+Click* dataset, which includes language-specific relevance scores for events and their relationships.

We experimentally demonstrate the advantage of employing language-specific embeddings for the task of language-specific event recommendation. *LaSER* consistently outperforms link-based, embedding-based and graph attention network-based recommendation baselines, achieving improvements of more than 8 and 17 percentage points in nDCG@10 and MAP@10, respectively. Moreover, a user study demonstrates that *LaSER* effectively recommends events within a language-specific context, outperforming the best-performing baselines by up to 33 percentage points in MAP@5, indicating that language-specific context is an important event recommendation criterion, alongside topical and global event relevance. The *Event-KG+Click* dataset emerges as a valuable resource for assessing event relevance within language-specific contexts. This dataset is part of the OEKG (Chap.6), effectively adopted by *LaSER*, and can potentially serve as a foundation for training and evaluating future cross-lingual event-centric analytics models. For example, understanding the relevance of an event in different languages can support the analysis of event-related news across the globe as performed in Chap. 10.

For future work, we intend to explore techniques for knowledge transfer from languages with rich resources to under-resourced languages, with the goal of extending *LaSER*'s applicability across languages not covered in the Wikipedia Clickstream.

Acknowledgements The work presented in this chapter has received funding from the European Union's Horizon 2020 research and innovation program under the Marie Skłodowska-Curie grant agreement no. 812997 and under the name CLEOPATRA (Cross-lingual Event-centric Open Analytics Research Academy).

References

Abdollahi S, Gottschalk S, Demidova E (2020) EventKG+Click: a dataset of language-specific event-centric user interaction traces. In: International Workshop on Cross-lingual Event-centric Open Analytics co-located with the Extended Semantic Web Conference (ESWC 2020), CEUR-WS.org, CEUR Workshop Proceedings, vol 2611, pp 32–42

Abdollahi S, Gottschalk S, Demidova E (2023) Laser: language-specific event recommendation. J Web Semant 75:100759. https://doi.org/10.1016/J.WEBSEM.2022.100759

AlGhamdi K, Shi M, Simperl E (2021) Learning to recommend items to wikidata editors. In: International Semantic Web Conference, Springer, pp 163–181. https://doi.org/10.1007/978-3-030-88361-4_10

Allan J, Papka R, Lavrenko V (1998) On-line new event detection and tracking. In: Annual International ACM SIGIR Conference on Research and Development in Information Retrieval, pp 37–45. https://doi.org/10.1145/3209978.3210136

Bi B, Ma H, Hsu BJ, Chu W, Wang K, Cho J (2015) Learning to recommend related entities to search users. In: ACM International Conference on Web Search and Data Mining, pp 139–148. https://doi.org/10.1145/2684822.2685304

Blanco R, Cambazoglu BB, Mika P, Torzec N (2013) Entity recommendations in web search. In: International Semantic Web Conference, Springer, pp 33–48. https://doi.org/10.1007/978-3-642-41338-4_3

Bordes A, Usunier N, Garcia-Duran A, Weston J, Yakhnenko O (2013) Translating embeddings for modeling multi-relational data. In: Advances in Neural Information Processing Systems 26

Burges CJ (2010) From RankNet to LambdaRank to LambdaMART: an overview. Learning 11(23–581):81

Cao S, Lu W, Xu Q (2016) Deep neural networks for learning graph representations. In: Schuurmans D, Wellman MP (eds) AAAI Conference on Artificial Intelligence, AAAI Press, pp 1145–1152

Gao L, Wu J, Qiao Z, Zhou C, Yang H, Hu Y (2016) Collaborative social group influence for event recommendation. In: ACM International Conference on Information and Knowledge Management, pp 1941–1944. https://doi.org/10.1145/2983323.2983879

Gottschalk S, Demidova E (2018) EventKG+ TL: creating cross-lingual timelines from an event-centric knowledge graph. In: European Semantic Web Conference, Springer, pp 164–169. https://doi.org/10.1007/978-3-319-98192-5_31

Gottschalk S, Bernacchi V, Rogers R, Demidova E (2018) Towards better understanding researcher strategies in cross-lingual event analytics. In: International Conference on Theory and Practice of Digital Libraries, Springer, pp 139–151. https://doi.org/10.1007/978-3-030-00066-0_12

Goyal P, Ferrara E (2018) Graph embedding techniques, applications, and performance: a survey. Knowl-Based Syst 151:78–94. https://doi.org/10.1016/j.knosys.2018.03.022

Grover A, Leskovec J (2016) Node2vec: scalable feature learning for networks. In: ACM SIGKDD International Conference on Knowledge Discovery and Data Mining, pp 855–864. https://doi.org/10.1145/2939672.2939754

Guo J, Fan Y, Pang L, Yang L, Ai Q, Zamani H, Wu C, Croft WB, Cheng X (2020a) A deep look into neural ranking models for information retrieval. Inf Process Manag 57(6):102067

Guo R, Zhao X, Henderson A, Hong L, Liu H (2020b) Debiasing grid-based product search in e-commerce. In: ACM SIGKDD International Conference on Knowledge Discovery & Data Mining, pp 2852–2860

Huang J, Ding S, Wang H, Liu T (2018) Learning to recommend related entities with serendipity for web search users. ACM Trans Asian Low-Resour Lang Inf Process 17(3):1–22. https://doi.org/10.1145/3185663

Järvelin K, Kekäläinen J (2002) Cumulated gain-based evaluation of IR techniques. ACM Trans Inf Syst 20(4):422–446. https://doi.org/10.1145/582415.582418

Khrouf H, Troncy R (2013) Hybrid event recommendation using linked data and user diversity. In: ACM Conference on Recommender Systems, pp 185–192. https://doi.org/10.1145/2507157.2507171

Kim D, Oh A (2021) How to find your friendly neighborhood: Graph attention design with self-supervision. In: International Conference on Learning Representations, ICLR, OpenReview.net

Liu JH, Goldstein-Hawes R, Hilton D, Huang LL, Gastardo-Conaco C, Dresler-Hawke E, Pittolo F, Hong YY, Ward C, Abraham S, et al (2005) Social representations of events and people in world history across 12 cultures. J Cross-Cult Psychol 36(2):171–191

Liu TY, et al (2009) Learning to rank for information retrieval. Found Trends® Inf Retr 3(3):225–331

Ni CC, Sum Liu K, Torzec N (2020) Layered graph embedding for entity recommendation using wikipedia in the yahoo! knowledge graph. In: The Web Conference, pp 811–818. https://doi.org/10.1145/3366424.3383570

Ou M, Cui P, Pei J, Zhang Z, Zhu W (2016) Asymmetric transitivity preserving graph embedding. In: ACM SIGKDD International Conference on Knowledge Discovery and Data Mining, pp 1105–1114. https://doi.org/10.1145/2939672.2939751

Palumbo E, Rizzo G, Troncy R (2017) Entity2rec: learning user-item relatedness from knowledge graphs for top-n item recommendation. In: ACM Conference on Recommender Systems, pp 32–36

Perozzi B, Al-Rfou R, Skiena S (2014) Deepwalk: online learning of social representations. In: ACM SIGKDD International Conference on Knowledge Discovery and Data Mining, pp 701–710. https://doi.org/10.1145/2623330.2623732

Qiao Z, Zhang P, Cao Y, Zhou C, Guo L, Fang B (2014) Combining heterogenous social and geographical information for event recommendation. In: Brodley CE, Stone P (eds) Proceedings of the Twenty-Eighth AAAI Conference on Artificial Intelligence, AAAI Press, pp 145–151. https://doi.org/10.1609/aaai.v28i1.8725

Rudnik C, Ehrhart T, Ferret O, Teyssou D, Troncy R, Tannier X (2019) Searching news articles using an event knowledge graph leveraged by wikidata. In: World Wide Web Conference, pp 1232–1239. https://doi.org/10.1145/3308560.3316761

Setty V, Hose K (2018) Event2vec: neural embeddings for news events. In: International ACM SIGIR Conference on Research & Development in Information Retrieval, pp 1013–1016

Toutanova K, Lin XV, Yih Wt, Poon H, Quirk C (2016) Compositional learning of embeddings for relation paths in knowledge base and text. In: Annual Meeting of the Association for Computational Linguistics (Volume 1: Long Papers), pp 1434–1444

Tran NK, Tran T, Niederée C (2017) Beyond time: dynamic context-aware entity recommendation. In: European Semantic Web Conference, Springer, pp 353–368. https://doi.org/10.1007/978-3-319-58068-5_22

Wang Q, Mao Z, Wang B, Guo L (2017) Knowledge graph embedding: a survey of approaches and applications. IEEE Trans Knowl Data Eng 29(12):2724–2743

Witten IH, Milne DN (2008) An effective, low-cost measure of semantic relatedness obtained from wikipedia links. In: AAAI Workshop on Wikipedia and Artificial Intelligence: An Evolving Synergy, AAAI Press, pp 25–30

Wu L, Hu D, Hong L, Liu H (2018) Turning clicks into purchases: revenue optimization for product search in e-commerce. In: International ACM SIGIR Conference on Research & Development in Information Retrieval, pp 365–374

Xie R, Liu Z, Sun M, et al. (2016) Representation learning of knowledge graphs with hierarchical types. In: IJCAI, vol 2016, pp 2965–2971

Yamada I, Asai A, Sakuma J, Shindo H, Takeda H, Takefuji Y, Matsumoto Y (2020) Wikipedia2vec: an efficient toolkit for learning and visualizing the embeddings of words and entities from wikipedia. In: Conference on Empirical Methods in Natural Language Processing: System Demonstrations, pp 23–30

Zhang L, Rettinger A, Zhang J (2016) A probabilistic model for time-aware entity recommendation. In: International Semantic Web Conference, Springer, pp 598–614. https://doi.org/10.1007/978-3-319-46523-4_36

Chapter 9
Conversational Question Answering over Knowledge Graphs

Endri Kacupaj, Kuldeep Singh, Maria Maleshkova, and Jens Lehmann

Abstract Question answering (QA) over knowledge graphs (KGs) is an essential task that maps a user's utterance to a query over a KG to retrieve the correct answer. Earlier methods in this field relied heavily on predefined templates and rules, which had limited adaptability and learning capability. Recent research has made significant strides in answering straightforward questions, and there has been notable success in tackling more intricate queries as well. However, a key challenge remains that, often, a single round of question and answer is not enough. Users might have follow-up questions that delve deeper into a subject, relying on information from their initial queries. This situation is particularly common in conversational settings, where each new question might refer back to earlier topics or answers. In this chapter, we explore advanced techniques for handling such conversational QA over knowledge graphs. We leverage deep neural networks and multi-task learning approaches to create systems that can understand and respond to a series of interconnected questions. By focusing on these conversational aspects and the nuances between different types of queries, we aim to bridge a significant gap in current research, offering more dynamic and context-aware systems that can adapt to the evolving nature of human inquiry.

E. Kacupaj (✉)
Cerence GmbH, Bonn, Germany
e-mail: endri.kacupaj@cerence.com

K. Singh
Zerotha Research and Cerence GmbH, Aachen, Germany
e-mail: kuldeep.singh1@cerence.com

M. Maleshkova
Helmut-Schmidt-Universität/Universität der Bundeswehr Hamburg, Hamburg, Germany
e-mail: maria@maleshkova.com

J. Lehmann
TU Dresden & Amazon, Dresden, Germany
e-mail: jens.lehmann@tu-dresden.de

I. Marenzi et al. (eds.), *Event Analytics across Languages and Communities*,
https://doi.org/10.1007/978-3-031-64451-1_9

9.1 Introduction

Question-answering (QA) interfaces have become a popular method for querying information in various formats, like unstructured (e.g. news articles) and structured (e.g. knowledge graphs), using natural language. Knowledge graphs (KGs) have attracted significant interest from both industry and academia, as they curate and connect facts from diverse Web-based information sources. Large-scale KGs such as DBpedia (Lehmann et al. 2015) and Wikidata (Vrandečić and Krötzsch 2014) are publicly available and widely used as reference sources of information and knowledge in various fields (Chap. 6; Chapter 2) (Gottschalk et al. 2021; Tahmasebzadeh et al. 2021; Guluzade et al. 2021), including information retrieval and question answering.

Knowledge graph question-answering (KGQA) systems aim to map a user's natural language question to a query over a KG to retrieve the correct answer. KGQA approaches have gained considerable attention in recent years, and a large number of methods have been proposed (Lan et al. 2021).

The emergence of intelligent personal assistants, such as Alexa,[1] Siri[2] and Google Assistant,[3] has shifted the focus of research towards conversational question-answering (ConvQA) or multi-turn QA systems. These systems aim to understand the context of the given query and engage users in a conversation to satisfy their information needs. In recent years, ConvQA over knowledge graphs (KGs) has gained attention and prominence due to the availability of large-scale multi-turn QA datasets and advancements in the field of deep learning. Figure 9.1 illustrates an example of ConvQA over KGs.

Numerous studies (Singh 2019; Dubey 2021; Zafartavanaelmi 2021) have tackled the KGQA task by creating stand-alone components. However, this may result in limited information sharing, causing the components to work independently and possibly overlook valuable information that could improve QA performance. To overcome this, adopting a multi-task learning (MTL) approach to jointly train related tasks and share information, such as representations, can be beneficial. This enables the broader QA architecture to generalise better on the overall task. MTL has become a prominent learning paradigm in various fields (Ruder 2017) and is the primary approach employed in the work presented in this chapter.

While our focus has been on developing general-purpose ConvQA systems, they can also be tailored for specific applications. For instance, these systems can be trained on event-centric, multilingual information to allow users to ask questions and receive accurate, relevant answers in various languages. In situations like terrorist attacks or natural disasters, a ConvQA system can be trained to extract event-centric, multilingual data from sources such as news articles, social media

[1] https://developer.amazon.com/en-US/alexa

[2] https://www.apple.com/siri/

[3] https://assistant.google.com/

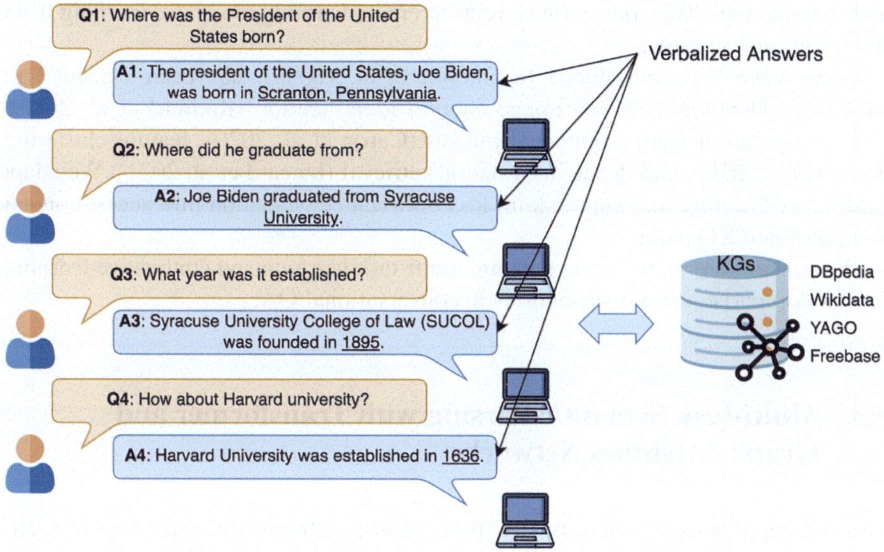

Fig. 9.1 A conversational question-answering example with answers derived from knowledge graphs (Figure from Kacupaj 2022)

and official reports. Users can then ask questions in multiple languages about the incident, like "What caused the explosion?" or "How many people were injured?" The ConvQA system can provide answers in the user's preferred language by processing information from the knowledge graph. Thus, ConvQA systems present a valuable solution for managing and utilising event-centric, multilingual information by allowing users to ask complex queries in natural language and receive accurate, relevant answers in different languages.

9.2 Related Work

Conversational AI has seen various neural approaches, as surveyed by Gao et al. (2018). We focus on related work in semantic parsing, contrastive learning and multi-task learning in conversations while briefly mentioning other approaches.

Key studies include (Liang et al. 2017) with a neural symbolic machine (NSM) and key-value memory network; (Saha et al. 2018) proposing a hybrid Hierarchical Recurrent Encoder-Decoder model (HRED) and key-value memory network model; (Guo et al. 2018) presenting a model converting conversational utterances to logical forms; (Shen et al. 2019) introducing a multi-task learning framework for type-aware entity detection and logical form generation; (Christmann et al. 2019) proposing a graph exploration algorithm for conversational questions over a KG;

and (Kaiser et al. 2021) presenting a reinforcement learning model for learning from conversational streams.

Contrastive loss, introduced in Chopra et al. (2005), has been extended to various applications, such as image caption identification (Radford et al. 2019), code computation from augmented images (Caron et al. 2020), feature clustering (Caron et al. 2018) and dense information retrieval (Izacard et al. 2021). We adapt contrastive learning to compute joint loss between conversation utterances, context and candidate KG paths.

We concentrate on semantic parsing, multi-task learning and contrastive learning for our comparison and contributions in conversational QA.

9.3 Multi-task Semantic Parsing with Transformer and Graph Attention Networks

The first approach we present is a multi-task learning framework called LASAGNE (Kacupaj et al. 2021b) that combines a transformer model with graph attention networks (GATs) for neural semantic parsing across multiple tasks (Veličković et al. 2018). The approach includes tasks such as named entity recognition, entity linking, relation extraction and logical form generation. Our approach incorporates semantic parsing with a transformer model (Vaswani et al. 2017) in a manner similar to prior work. However, LASAGNE distinguishes itself through two key innovations: (1) We enhance the transformer model with a graph attention network to take advantage of the relationships between entity types and predicates, benefiting from the message-passing ability between nodes. (2) We develop a new entity recognition module that identifies, links, filters and rearranges all pertinent entities. Our experimental results demonstrate that these innovative contributions lead to significant performance improvements. LASAGNE achieves state-of-the-art results in eight out of ten question types on the Complex Sequential Question Answering (CSQA) dataset (Saha et al. 2018).

9.3.1 Approach

The input data in a conversation comprises utterances u and corresponding answers a from the knowledge graph. LASAGNE, our framework, uses multi-task semantic parsing to map utterance u to logical form z based on the conversation context, as shown in Fig. 9.2.

For semantic parsing, we use a grammar that contains the minimum number of actions while capturing the input utterance context. We integrate most actions from Plepi et al. (2021) and update some for improved performance. To convert the input conversation into a sequence of actions (logical form), a transformer model

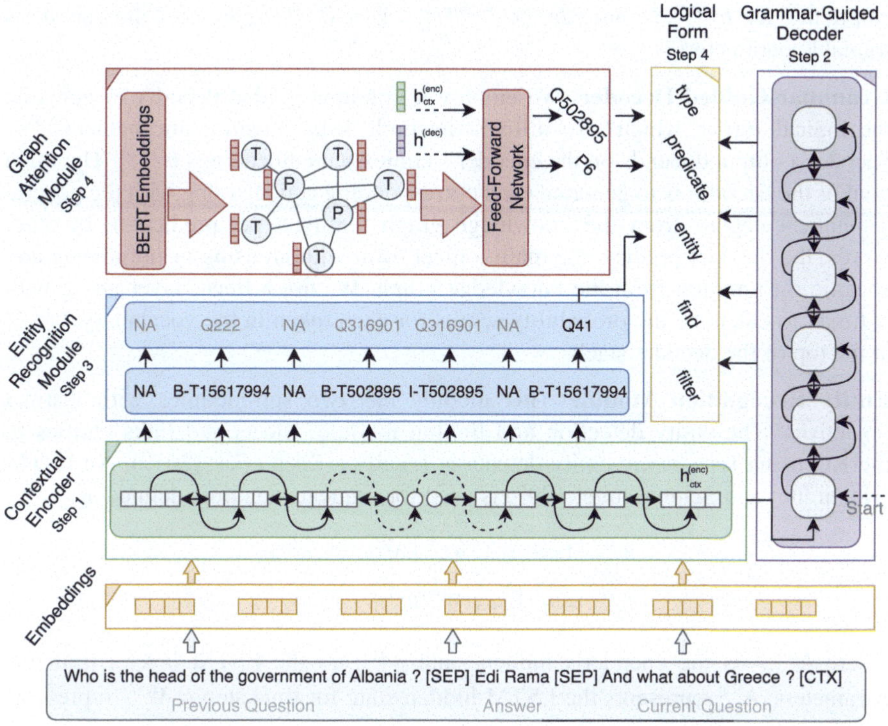

Fig. 9.2 The LASAGNE architecture blends transformer and graph attention network technologies for semantic parsing. It has three key components. Firstly, a transformer model with a contextual encoder and a grammar-guided decoder. Secondly, an entity recognition module that identifies and classifies entities from the context, linking them to a knowledge graph and adjusting their arrangement as needed. Lastly, a graph attention module using a GAT network with BERT embeddings, which analyses relationships between entity types and predicates. This module combines node embeddings with context and decoder hidden states to accurately predict node types and predicates (Figure from Kacupaj et al. 2021b)

(Vaswani et al. 2017) is employed. It maps a question q as sequence x to label l, a sequence y, by modelling conditional probability $p(y|x)$.

Input and Word Embedding The model incorporates dialogue history from previous interactions to handle co-reference and ellipsis by considering the previous question, answer and current question for each turn as utterances, separated by the $[SEP]$ token. A $[CTX]$ token is appended at the end to represent the entire input. The conversation context is tokenised using WordPiece tokenisation (Wu et al. 2016). Pre-trained GloVe model (Pennington et al. 2014) is then used to embed words into a vector space of dimension d.

Contextual Encoder The word embeddings x is passed as input to the contextual encoder, which uses the multi-head attention mechanism introduced by Vaswani et al. (2017). The encoder outputs the contextual embeddings $h^{(enc)}$, which we can define as: $h^{(enc)} = encoder(x; \theta^{(enc)})$, where $\theta^{(enc)}$ represents the encoder's trainable parameters.

Grammar-Guided Decoder We employ a grammar-guided decoder to generate the logical forms, which also utilises the multi-head attention mechanism. The decoder's output depends on the encoder's contextual embeddings $h^{(enc)}$. The main goal of the decoder is to generate each corresponding action, along with the general semantic category from the knowledge graph (entity, type, predicate). In other words, the decoder predicts the main logical form without using or initialising any specific information from the knowledge graph. We use a linear layer along with softmax to calculate the probability scores for each token in the vocabulary $V^{(dec)}$ at the top of the decoder stack.

Entity Recognition Module This module has two sub-modules with distinct objectives. The entity detection and linking module detects and links entities to the KG using type-aware entity detection, based on Shen et al. (2019). An LSTM (Hochreiter and Schmidhuber 1997) is used for sequence tagging, defined as:

$$\begin{aligned} h^{(l)} &= \text{LeakyReLU}(LSTM(h^{(enc)}; \theta^{(l)})), \\ p_t^{(ed)} &= softmax(\mathbf{W}^{(l)} h_t^{(l)}), \end{aligned} \tag{9.1}$$

where $h^{(enc)}$ is the encoder's hidden state, $\theta^{(l)}$ are the LSTM layer's trainable parameters, $h_t^{(l)}$ represents the LSTM hidden state for time step t, $\mathbf{W}^{(l)}$ represents the linear layer weights and $p_t^{(ed)}$ is the entity detection module's prediction at time step t. After the entity detection, the entity linking process has three steps: (1) identify entity spans, (2) create an inverted index for KG entities and (3) filter retrieved candidates using predicted types.

The filtering and permutation module takes $h^{(enc)}$ and $h^{(l)}$ as input and assigns index tags using a feed-forward network.

Graph Attention-Based Module To construct the graph, we examine relations and entity types in the KG. We define graph $\mathcal{G}$ as the set of types, relations and links with existing triples $(e_1, r, e_2) \in \mathcal{K}$, where e_1 has type tp_1, e_2 has type tp_2 and r is a relation. We employ GATs (Veličković et al. 2018) to propagate information and project KG data into the embedding space. Several works employ graph-based networks in different domains (Plepi and Flek 2021; Plepi et al. 2022). Node embeddings $h^{(g)}$ are initialised with pre-trained BERT (Devlin et al. 2019) embeddings. A GAT layer transforms input representations as: $\overline{h}^{(g)} = g(h^{(g)}; \theta^{(g)})$, with $\theta^{(g)}$ as trainable parameters. Predicting the correct type or predicate in the logical form is modelled as a classification task over the nodes in graph $\mathcal{G}$. For each decoder time step t, we compute the probability distribution $p_t^{(g)}$ over graph nodes as: $p_t^{(g)} = softmax(\overline{h}^{(g)T} h_t^{(c)})$, where $h_t^{(c)}$ is a linear projection of the context representation and decoder hidden state concatenation: $h_t^{(c)} = \text{LeakyReLU}(\mathbf{W}^{(g)}[h_{ctx}^{(enc)}; h_t^{(dec)}])$, with $\mathbf{W}^{(g)} \in \mathbb{R}^{d \times 2d}$.

9.3.2 Multi-task Learning

The proposed framework consists of four trainable modules: a grammar-guided decoder, entity detection, filtering and permutation and a GAT-based module for types and predicates. Each module contributes to the overall performance with a loss function similar to other works (Armitage et al. 2020). A weighted average of individual negative log-likelihood losses is used for multi-tasking: $L = \lambda_1 L^{dec} + \lambda_2 L^{ed} + \lambda_3 L^{ef} + \lambda_4 L^{g}$, where λ_1, λ_2, λ_3, and λ_4 are learned weights.

9.3.3 Experimental Setup

This subsection discusses the experimental setup, including datasets, resources and metrics used.

Datasets Experiments are conducted on the Complex Sequential Question Answering (CSQA) dataset[4] (Saha et al. 2018). The CSQA dataset is created using the Wikidata knowledge graph, containing 21.2 million triples with over 12.8 million entities, 3,054 entity types and 567 predicates. It comprises around 200,000 dialogues, with train, validation and test partitions containing 153,000, 16,000 and 28,000 dialogues, respectively. The questions require complex reasoning to find the correct answers.

Model Configurations The CSQA dataset is annotated with gold logical forms following a similar approach to previous work (Plepi et al. 2021). The transformer module is configured according to (Vaswani et al. 2017), with $d = 300$, $H = 6$ heads, $L = 2$ layers and $d_{ff} = 600$. A residual dropout of 0.1 is applied in both encoder and decoder stacks. The entity detection and filtering & permutation module have a 300-dimensional output. The GAT-based module utilises pre-trained BERT embeddings, with an input size of 3072. The GAT layer generates 300-dimensional representations. Optimisation is done using the Noam optimiser, which uses an Adam optimiser (Kingma and Ba 2015) and warm-up steps for the learning rate.

Models for Comparison The LASAGNE framework is compared to three baselines evaluated on the CSQA dataset: HRED+KVmem model (Saha et al. 2018), D2A (Guo et al. 2018) and MaSP (Shen et al. 2019).

Evaluation Metrics The evaluation metrics used are the same as those employed by the authors of the CSQA dataset (Saha et al. 2018) and previous baselines. F1-score is used for questions with answers composed of a set of entities, while the Accuracy metric is used for quantitative questions or Boolean value (YES/NO)

[4] https://amritasaha1812.github.io/CSQA

Table 9.1 LASAGNE's performance on the CSQA dataset

Methods		HRED-KVM	D2A	MaSP	LASAGNE	
# Train param		–	–	15M	14.7M	Δ
Question type	#Examples	F1 score				
Overall	206k	9.39%	66.70%	79.26%	**82.91%**	+3.65%
Clarification	12k	16.35%	35.53%	**80.79%**	69.46%	−11.33%
Comparative reasoning (All)	15k	2.96%	48.85%	68.90%	**69.77%**	+0.87%
Logical reasoning (All)	22k	8.33%	67.31%	69.04%	**89.83%**	+20.79%
Quantitative reasoning (All)	9k	0.96%	56.41%	73.75%	**86.67%**	+12.92%
Simple question (coreferenced)	55k	7.26%	57.69%	76.47%	**79.06%**	+2.59%
Simple question (direct)	82k	13.64%	78.42%	85.18%	**87.95%**	+2.77%
Simple question (ellipsis)	10k	9.95%	81.14%	**83.73%**	80.09%	−3.64%
Question type	#Examples	Accuracy				
Overall	66k	14.95%	37.33%	45.56%	**64.34%**	+18.78%
Verification (boolean)	27k	21.04%	45.05%	60.63%	**78.86%**	+18.23%
Quantitative reasoning (count)	24k	12.13%	40.94%	43.39%	**55.18%**	+11.79%
Comparative reasoning (count)	15k	8.67%	17.78%	22.26%	**53.34%**	+31.08%

Best results are denoted in bold

answers. An overall score for each evaluation metric and their corresponding question categories is also provided.

9.3.4 Results

Table 9.1 compares LASAGNE with earlier baselines, revealing its superior performance in overall weighted average and in eight of ten question types, with gains up to 31%. For multi-entity reasoning questions like *Logical Reasoning (All)* and *Verification (Boolean)*, LASAGNE's entity recognition module leads to significant improvements (+20.79% and +18.23%). It also outperforms MaSP in *Quantitative Reasoning* and *Comparative Reasoning* categories due to its graph attention-based module. LASAGNE performs better in two out of three *Simple Question* cases, displaying its robustness. Nonetheless, it underperforms in *Clarification* questions due to spurious logical forms, which also affect *Simple Questions (Ellipsis)* performance.

Table 9.2 The effectiveness of the GAT-based module and multi-task learning in LASAGNE

Methods	Ours	w/o GATs	w/o MTL
Question type	F1 score		
Clarification	66.94%	57.33%	59.43%
Comparative	69.77%	57.72%	66.41%
Logical	89.83%	78.52%	86.75%
Quantitative	86.67%	75.26%	82.18%
Simple (Coref)	79.06%	76.46%	77.23%
Simple (Direct)	87.95%	83.59%	85.39%
Simple (Ellipsis)	80.09%	77.19%	78.47%
Question type	Accuracy		
Verification	78.86%	63.38%	75.24%
Quantitative	55.18%	40.87%	46.27%
Comparative	53.34%	41.73%	45.90%

9.3.5 Ablation Study and Error Analysis

Ablation Study Table 9.2 emphasises the benefits of using the GAT-based module and multi-task learning in LASAGNE. Replacing the GAT-based module with simpler classifiers results in decreased performance for question types needing multiple entity types and predicates. Removing multi-task learning and training modules separately negatively affects all question types, as LASAGNE's filtering & permutation and GAT-based modules rely on supervision signals from previous modules. Without multi-task learning, each module must re-learn inherited information, leading to underperformance.

Error Analysis In our error analysis of 100 random incorrect predictions, we found two main issues: (1) Entity ambiguity, where the framework struggles to link entities with the same name and type, even when correctly identifying them. For example, "Jeff Smith" as a "common name" could refer to multiple entities in Wikidata. (2) Lack of gold actions for some question categories, leading to inaccurate logical forms, particularly in "Comparative, Quantitative and Clarification" categories. Despite this, our model achieved state-of-the-art performance in comparative and quantitative aspects.

9.4 Knowledge Graph Path Ranking via Contrastive Representation Learning

In this section, we describe PRALINE (Kacupaj et al. 2022b), a novel ConvQA method for KGs utilising path ranking and fluent answer verbalisation. It consists of four concurrently trained modules: input conversation encoding, domain identification, joint conversation and KG-path embedding and answer verbalisation for enhanced KG-path ranking context.

9.4.1 Approach

In a conversation, the input data consists of questions q^t and answers a^t extracted from the knowledge graph. We propose a contrastive learning approach called PRALINE to rank KG context paths $\mathcal{P}_c^t$ based on conversations. Figure 9.3 demonstrates PRALINE's operation, which involves three steps: (1) pre-processing and extracting potential candidate paths (and their domains), (2) encoding conversational context while using fluent verbalised responses (Kacupaj et al. 2020, 2021a,c) that provide sufficient information to enrich the learning process and (3) jointly embedding the conversation, its context and candidate KG paths in a common space to apply a contrastive ranking module and effectively rank KG paths.

Preprocessing and Extracting Representations First, we identify the context entities $\mathcal{E}_c$ and extract potential candidates for KG paths, as done by Kaiser et al. (2021). Next, we extract the KG paths $\mathcal{P}_c$ and initialise their representations using sentence embeddings derived from a BERT model (Devlin et al. 2019). Each KG path is considered a sentence and input to BERT. We use these KG path representations for the contrastive ranking task. Additionally, we pre-process and embed the conversation domains to obtain the representation $h^{(dm)}$. These embedded domains are implicitly used during the ranking process.

Encoding Conversation History and Question We use a BART-based bidirectional encoder (Lewis et al. 2020) to encode conversation history C^t and current question q^t. The input sequence s^t is created by concatenating the conversation history and question using a $[SEP]$ token. We tokenise s^t and pass it through the encoder, producing contextual embeddings $h^{(enc)}$. A BART-based decoder generates response v^t, with its vocabulary $V^{(dec)}$ containing all unique tokens and a helper token for answer position a^t. A linear layer and softmax calculate the probability score for each token in the decoder vocabulary.

Domain Identification Pointer PRALINE's second step uses a domain identification pointer network to identify the KG domain of the input sequence s^t. This pointer network is designed to handle different sizes of vocabulary and allows for updates to the domain vocabulary, which is represented as $V^{(dm)}$. To calculate pointer scores, we use the contextual embeddings from the encoder, denoted as $h^{(enc)}$. The pointer network includes a simple linear network and a softmax layer. The pointer scores are given by the equation $\omega_i^{(dm)} = \text{softmax}(W_1^{(dm)} u_t^{(dm)})$, where $W_1^{(dm)}$ is a set of weights used in this calculation. We also compute a joint representation $u_t^{(dm)}$ that combines domain and contextual embeddings. This is calculated with the formula $u_t^{(dm)} = \tanh(W_2^{(dm)}\tau + h^{(enc)})$. Here, $W_2^{(dm)}$ is another set of weights, and τ represents the domain embeddings. The dimensions of these embeddings

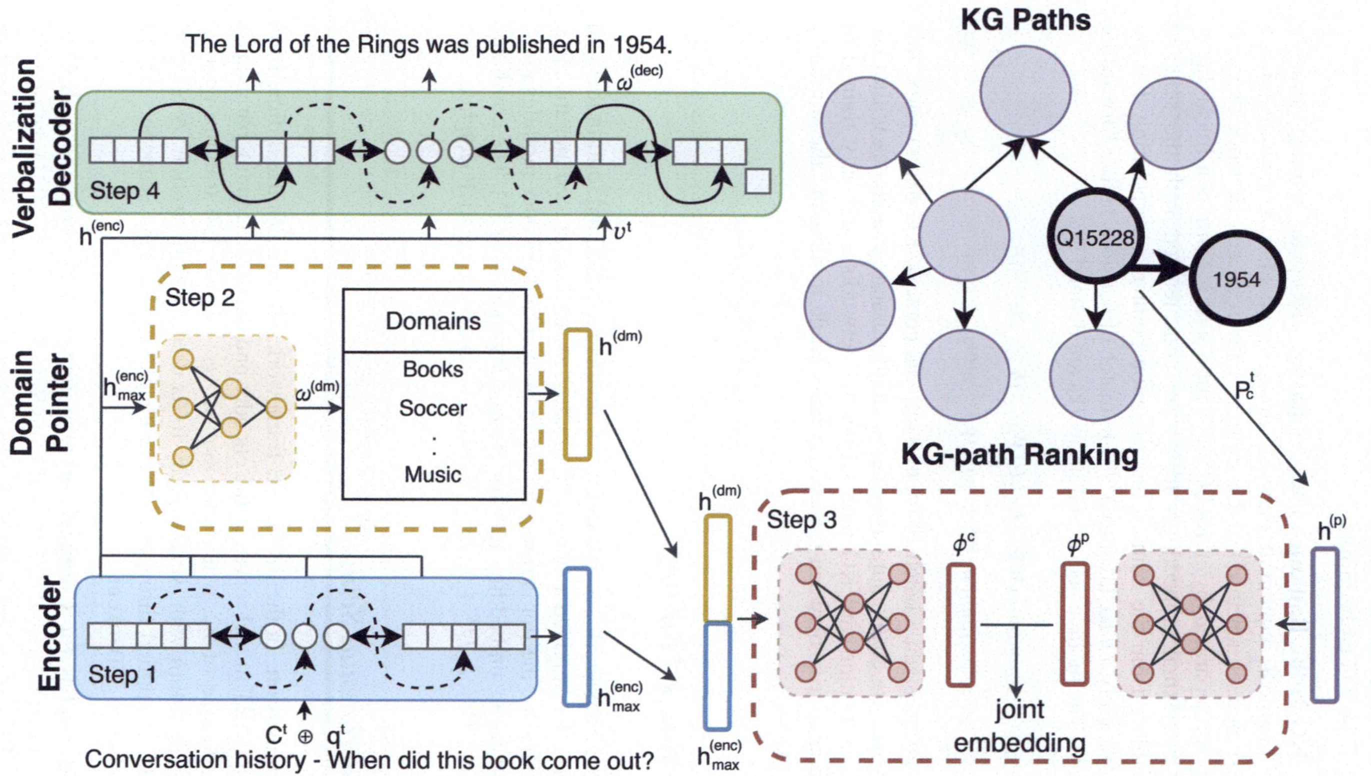

Fig. 9.3 The PRALINE architecture, designed for conversational question answering, consists of three main steps. First, it extracts knowledge graph paths and domains, representing them using a BERT model. Second, it learns the conversational context using a BART model, augmented with a domain identification pointer. The third step involves a contrastive ranking module. This module develops a joint embedding space for the conversation, incorporating both contextual embeddings from the encoder and selected domain embeddings, as well as for the context path (Figure from Kacupaj 2022)

and weights are adjusted to match the size of the domain vocabulary and the requirements of the KG embeddings.

Contrastive KG-Path Ranking The contrastive KG-path ranking module employs two identical sequential networks to generate joint embeddings for a conversation (input sequence s^t) and a context path p_c^t. Each network contains two linear layers separated by a $ReLU$ activation and a $tanh$ non-linear layer. The encoder contextual embeddings $h^{(enc)}$ are concatenated with the selected domain embedding from the domain identification pointer, incorporating domain information. We apply a max layer to $h^{(enc)}$ to match dimensions before forwarding it to the module. The module sequential networks are defined as:

$$\begin{aligned}\phi^c &= tanh(\mathbf{W}_2^{(crk)} ReLU(\mathbf{W}_1^{(crk)}[h_{max}^{(enc)}; h^{(dm)}])),\\ \phi^p &= tanh(\mathbf{W}_2^{(prk)} ReLU(\mathbf{W}_1^{(prk)} h^{(p)})),\end{aligned} \tag{9.2}$$

During training, given a batch of (conversational context, KG paths) pairs, this module computes the cosine similarity between all possible candidates within the batch. The conversation and KG path sequential networks are jointly trained to maximise the similarity between the correct pairs while minimising the similarity for incorrect pairs.

9.4.2 Multi-task Learning

The PRALINE framework consists of three trainable modules, and to train them all simultaneously, the weighted average of individual losses is computed: negative log-likelihood losses for the domain identification pointer and decoder modules (L^{dm} and L^{dec}) and cosine embedding loss for the ranking module (L^{rk}). The combined loss is $L = \lambda_1 L^{dm} + \lambda_2 L^{rk} + \lambda_3 L^{dec}$, with λ_1, λ_2 and λ_3 as relative weights.

9.4.3 Experimental Setup

Model Configuration In the PRALINE framework, all modules use a dimension $d = 768$. The BART model serves as the encoder and decoder. Training parameters include a batch size of 32, a learning rate of $1e - 4$ and 120 training epochs. The AdamW algorithm with weight decay fix is used for optimisation. A residual dropout of 0.1 is applied throughout the framework. The input sequence size of $C^t + q^t$ is limited to 150 tokens.

Datasets and Models for Comparison Our framework is compared with ConvQuestions (Christmann et al. 2019) and ConvRef (Kaiser et al. 2021) datasets for ConvQA over KGs, using fluent responses (Kacupaj et al. 2022a). The first baseline, CONVEX (Christmann et al. 2019), is a two-stage process for detecting KG answers to conversational utterances. The second baseline, CONQUER (Kaiser et al. 2021), is the current state-of-the-art RL-based method for conversational QA over KGs. OAT (Marion et al. 2021), a semantic parsing-based approach, and Focal entity (Lan and Jiang 2021) are other recent models. For ConvQuestions, baseline values come from the official leaderboard, while for ConvRef, values are from baseline papers.

Evaluation Metrics To evaluate ConvQA performance, we use the following ranking metrics, also employed by previous baselines: (i) Precision at the top rank (P@1); (ii) Mean Reciprocal Rank (MRR), the average across the reciprocal of the rank at which the first context path was retrieved; and (iii) Hit at 5 (H@5), the fraction of times a correct answer was retrieved within the top-5 positions.

9.4.4 Results

Table 9.3 summarises the results, comparing PRALINE against previous baselines. On the ConvQuestions dataset, PRALINE outperforms baselines in all metrics. Specifically, PRALINE's P@1 outperforms CONQUER, CONVEX and OAT by 0.052, 0.108 and 0.042 points, respectively. The margins for H@5 and MRR are even more significant, with PRALINE beating CONQUER by 0.186 and 0.119 points, respectively, and CONVEX by 0.310 and 0.198 points. PRALINE outperforms OAT by 0.138 in MRR.

On the ConvRef dataset, PRALINE outperforms CONVEX in all metrics with a margin of over 0.100 absolute points. It also surpasses CONQUER in H@5 and MRR by 0.170 and 0.046 points, respectively. ConvRef extends ConvQuestions with multiple question reformulations, and CONQUER leverages these to improve results. PRALINE, however, treats reformulated questions the same as the original questions from the ConvQuestions benchmark, leading to a less significant P@1 increase.

Table 9.3 Overall results on employed datasets

Dataset	ConvQuestions			ConvRef		
Model	P@1	H@5	MRR	P@1	H@5	MRR
CONVEX	0.184	0.219	0.200	0.225	0.257	0.241
CONQUER	0.240	0.343	0.279	**0.353**	0.429	0.387
OAT	0.250	–	0.260	–	–	–
Focal entity	0.248	–	0.248	–	–	–
PRALINE	**0.292**	**0.529**	**0.398**	0.335	**0.599**	**0.441**

Best results are denoted in bold

Table 9.4 Ablation study

Dataset	ConvQuestions			ConvRef		
Model	P@1	H@5	MRR	P@1	H@5	MRR
PRALINE	0.292	0.529	0.398	0.335	0.599	0.441
w/o full conv.	0.214	0.375	0.299	0.247	0.449	0.324
w/o domain	0.247	0.436	0.296	0.266	0.472	0.356
w/o fluent resp.	0.265	0.441	0.324	0.279	0.503	0.397
Train separately	0.255	0.413	0.328	0.304	0.529	0.408

9.4.5 Ablation Study and Error Analysis

Ablation Study Table 9.4 illustrates various ablation studies performed on the PRALINE approach and its related architecture choices. The experiments include analysing the effect of full conversational history, domain information, fluent response and joint training on the performance of the approach. The results indicate that conversational contexts positively impact KG path ranking, and joint training of modules is effective. The results justify the use of various modules in PRALINE and support its overall performance.

Error Analysis We analysed 250 random incorrect predictions and found two error types: (1) PRALINE often misranks paths with semantically similar relations, such as "genre (P136)" and "main subject (P921)". (2) Over 25% of training examples and 19% of test examples lack gold paths, affecting the learning process and results. Improved annotation of gold KG paths could enhance PRALINE's performance.

9.5 Conclusion

In conclusion, this chapter has provided a comprehensive examination of two distinct methodologies for conversational question answering over knowledge graphs. These approaches have been meticulously explored in order to elucidate their respective strengths and potential applications in the field of natural language processing and artificial intelligence.

The first method utilises a multi-task learning paradigm to generate logical forms, which are then executed over the knowledge graph to obtain accurate responses. This technique leverages the inherent synergies between various tasks, consequently improving the performance of the model by learning shared representations. Multi-task learning has been demonstrated to be highly effective in addressing complex conversational scenarios, as it enables the model to effectively disentangle and contextualise information from different sources.

The second approach, on the other hand, focuses on information retrieval through the ranking of knowledge graph paths based on conversational history.

This technique capitalises on the rich contextual information provided by prior dialogue exchanges, enabling the model to generate more relevant and coherent responses. By incorporating conversational history into the ranking process, this method effectively enhances the overall performance of the question-answering system, delivering more accurate and contextually appropriate responses.

In summary, the detailed examination of these two approaches has yielded valuable insights into the nuances and complexities of conversational question answering over knowledge graphs. By understanding the intricacies of these methodologies, researchers and practitioners alike can make informed decisions when developing and implementing question-answering systems in various domains. Furthermore, the exploration of these techniques sets the stage for future research, fostering innovation and contributing to the advancement of the field. One key element that needs immediate attention from the research community is how to leverage the power of large language models (LLMs) for conversational question answering. Moreover, exploration of LLMs for this task requires consideration of ethics, latency and credibility of answers. Future work in this area includes the utilisation of event-based knowledge graphs such as EventKG (Gottschalk and Demidova 2018) or OEKG (Gottschalk et al. 2021), which can offer rich, context-specific data that is crucial for enhancing the effectiveness of question-answering systems in dynamic and diverse information environments.

Acknowledgments The work presented in this chapter has received funding from the European Union's Horizon 2020 research and innovation program under the Marie Skłodowska-Curie grant agreement no. 812997 and under the name CLEOPATRA (Cross-lingual Event-centric Open Analytics Research Academy).

References

Armitage J, Kacupaj E, Tahmasebzadeh G, Maleshkova M, Ewerth R, Lehmann J (2020) MLM: a benchmark dataset for multitask learning with multiple languages and modalities. In: Proceedings of the 29th ACM international conference on information & knowledge management, pp 2967–2974

Caron M, Bojanowski P, Joulin A, Douze M (2018) Deep clustering for unsupervised learning of visual features. In: Proceedings of the European conference on computer vision (ECCV), pp 132–149

Caron M, Misra I, Mairal J, Goyal P, Bojanowski P, Joulin A (2020) Unsupervised learning of visual features by contrasting cluster assignments. Adv Neural Inf Proces Syst 33:9912–9924

Chopra S, Hadsell R, LeCun Y (2005) Learning a similarity metric discriminatively, with application to face verification. In: 2005 IEEE computer society conference on computer vision and pattern recognition (CVPR'05), vol 1. IEEE, Piscataway, pp 539–546

Christmann P, Saha Roy R, Abujabal A, Singh J, Weikum G (2019) Look before you hop: conversational question answering over knowledge graphs using judicious context expansion. In: Proceedings of the 28th ACM international conference on information and knowledge management, pp 729–738

Devlin J, Chang MW, Lee K, Toutanova K (2019) Bert: pre-training of deep bidirectional transformers for language understanding. In: Proceedings of the 2019 conference of the North American chapter of the association for computational linguistics: human language technologies, vol 1 (long and short papers), pp 4171–4186

Dubey M (2021) Towards complex question answering over knowledge graphs. PhD thesis, University of Bonn

Gao J, Galley M, Li L (2018) Neural approaches to conversational AI. In: The 41st international ACM SIGIR conference on research & development in information retrieval, pp 1371–1374

Gottschalk S, Demidova E (2018) EventKG: a multilingual event-centric temporal knowledge graph. In: The semantic web: 15th international conference, ESWC 2018, Heraklion, June 3–7, 2018, Proceedings 15. Springer, Berlin, pp 272–287

Gottschalk S, Kacupaj E, Abdollahi S, Alves D, Amaral G, Koutsiana E, Kuculo T, Major D, Mello C, Cheema GS, et al. (2021) OEKG: The open event knowledge graph. In: CLEOPATRA@ WWW, pp 61–75

Guluzade A, Kacupaj E, Maleshkova M (2021) Demographic aware probabilistic medical knowledge graph embeddings of electronic medical records. In: International conference on artificial intelligence in medicine. Springer, Berlin, pp 408–417

Guo D, Tang D, Duan N, Zhou M, Yin J (2018) Dialog-to-action: conversational question answering over a large-scale knowledge base. In: Advances in neural information processing systems, pp 2942–2951

Hochreiter S, Schmidhuber J (1997) Long short-term memory. Neural Comput 9(8):1735–1780. https://doi.org/10.1162/NECO.1997.9.8.1735

Izacard G, Caron M, Hosseini L, Riedel S, Bojanowski P, Joulin A, Grave E (2021) Towards unsupervised dense information retrieval with contrastive learning. Preprint. arXiv:211209118

Kacupaj E (2022) Conversational question answering over knowledge graphs with answer verbalization. PhD thesis, Universitäts-und Landesbibliothek Bonn

Kacupaj E, Zafar H, Lehmann J, Maleshkova M (2020) VQuAnDa: verbalization question answering dataset. In: European semantic web conference. Springer, Berlin, pp 531–547

Kacupaj E, Banerjee B, Singh K, Lehmann J (2021a) ParaQA: a question answering dataset with paraphrase responses for single-turn conversation. In: European semantic web conference. Springer, Berlin, pp 598–613

Kacupaj E, Plepi J, Singh K, Thakkar H, Lehmann J, Maleshkova M (2021b) Conversational question answering over knowledge graphs with transformer and graph attention networks. In: Proceedings of the 16th conference of the European chapter of the association for computational linguistics: main volume, pp 850–862

Kacupaj E, Premnadh S, Singh K, Lehmann J, Maleshkova M (2021c) Vogue: answer verbalization through multi-task learning. In: Joint European conference on machine learning and knowledge discovery in databases. Springer, Berlin, pp 563–579

Kacupaj E, Singh K, Maleshkova M, Lehmann J (2022a) An answer verbalization dataset for conversational question answerings over knowledge graphs. Preprint. arXiv:220806734

Kacupaj E, Singh K, Maleshkova M, Lehmann J (2022b) Contrastive representation learning for conversational question answering over knowledge graphs. In: Proceedings of the 31st ACM international conference on information & knowledge management, pp 925–934

Kaiser M, Saha Roy R, Weikum G (2021) Reinforcement learning from reformulations in conversational question answering over knowledge graphs. In: 44th International ACM SIGIR conference on research and development in information retrieval. ACM, New York

Kingma DP, Ba J (2015) Adam: a method for stochastic optimization. In: Bengio Y, LeCun Y (eds) 3rd international conference on learning representations, ICLR 2015

Lan Y, Jiang J (2021) Modeling transitions of focal entities for conversational knowledge base question answering. In: Proceedings of the 59th annual meeting of the association for computational linguistics and the 11th international joint conference on natural language processing (vol 1: long papers)

Lan Y, He G, Jiang J, Jiang J, Zhao WX, Wen JR (2021) A survey on complex knowledge base question answering: methods, challenges and solutions. In: Proceedings of the thirtieth international joint conference on artificial intelligence, IJCAI-21, international joint conferences on artificial intelligence organization, pp 4483–4491. Survey track

Lehmann J, Isele R, Jakob M, Jentzsch A, Kontokostas D, Mendes PN, Hellmann S, Morsey M, Van Kleef P, Auer S, et al. (2015) DBpedia – a large-scale, multilingual knowledge base extracted from Wikipedia. Semant Web 6(2):167–195

Lewis M, Liu Y, Goyal N, Ghazvininejad M, Mohamed A, Levy O, Stoyanov V, Zettlemoyer L (2020) Bart: denoising sequence-to-sequence pre-training for natural language generation, translation, and comprehension. In: Proceedings of the 58th annual meeting of the association for computational linguistics, pp 7871–7880

Liang C, Berant J, Le Q, Forbus K, Lao N (2017) Neural symbolic machines: learning semantic parsers on freebase with weak supervision. In: Proceedings of the 55th annual meeting of the association for computational linguistics (vol 1: long papers), pp 23–33

Marion P, Nowak PK, Piccinno F (2021) Structured context and high-coverage grammar for conversational question answering over knowledge graphs. Preprint. arXiv:210900269

Pennington J, Socher R, Manning CD (2014) Glove: global vectors for word representation. In: Proceedings of the 2014 conference on empirical methods in natural language processing (EMNLP), pp 1532–1543

Plepi J, Flek L (2021) Perceived and intended sarcasm detection with graph attention networks. In: Findings of the association for computational linguistics: EMNLP 2021, pp 4746–4753

Plepi J, Kacupaj E, Singh K, Thakkar H, Lehmann J (2021) Context transformer with stacked pointer networks for conversational question answering over knowledge graphs. In: European semantic web conference. Springer, Berlin, pp 356–371

Plepi J, Sakketou F, Geiß HJ, Flek L (2022) Temporal graph analysis of misinformation spreaders in social media. Proceedings of textgraphs-16: graph-based methods for natural language processing, p. 89

Radford A, Wu J, Amodei D, Amodei D, Clark J, Brundage M, Sutskever I (2019) Better language models and their implications. OpenAI Blog 1:2. https://openai.com/research/better-language-models

Ruder S (2017) An overview of multi-task learning in deep neural networks. CoRR abs/1706.05098. https://doi.org/10.48550/arXiv.1706.05098

Saha A, Pahuja V, Khapra MM, Sankaranarayanan K, Chandar S (2018) Complex sequential question answering: towards learning to converse over linked question answer pairs with a knowledge graph. In: Proceedings of the thirty-second AAAI conference on artificial intelligence, (AAAI-18), the 30th innovative applications of artificial intelligence (IAAI-18), and the 8th AAAI symposium on educational advances in artificial intelligence (EAAI-18). AAAI Press, Washington, pp 705–713. https://doi.org/10.1609/AAAI.V32I1.11332

Shen T, Geng X, Qin T, Guo D, Tang D, Duan N, Long G, Jiang D (2019) Multi-task learning for conversational question answering over a large-scale knowledge base. In: Proceedings of the 2019 conference on empirical methods in natural language processing and the 9th international joint conference on natural language processing (EMNLP-IJCNLP), pp 2442–2451

Singh K (2019) Towards dynamic composition of question answering pipelines. PhD thesis, University of Bonn

Tahmasebzadeh G, Kacupaj E, Müller-Budack E, Hakimov S, Lehmann J, Ewerth R (2021) Geowine: geolocation based wiki, image, news and event retrieval. In: Proceedings of the 44th international ACM SIGIR conference on research and development in information retrieval, pp 2565–2569

Vaswani A, Shazeer N, Parmar N, Uszkoreit J, Jones L, Gomez AN, Kaiser Ł, Polosukhin I (2017) Attention is all you need. In: Advances in neural information processing systems, pp 5998–6008

Veličković P, Cucurull G, Casanova A, Romero A, Liò P, Bengio Y (2018) Graph attention networks. In: International conference on learning representations

Vrandečić D, Krötzsch M (2014) Wikidata: a free collaborative knowledgebase. Commun ACM 57(10):78–85
Wu Y, Schuster M, Chen Z, Le QV, Norouzi M, Macherey W, Krikun M, Cao Y, Gao Q, Macherey K, et al. (2016) Google's neural machine translation system: bridging the gap between human and machine translation. Preprint. arXiv:160908144
Zafartavanaelmi H (2021) Semantic question answering over knowledge graphs: pitfalls and pearls. PhD thesis, University of Bonn

Part III
Event Analytics

The chapters in Part III of this volume explore the dissemination and communication of events online across different languages, cultures and geographical areas. They are all in different ways concerned with telling stories about and with multilingual data. Addressing the challenges of working across diverse languages and national contexts is key to developing shared understandings of the events that shape the world around us and to building the critical data literacy skills that will help individuals and societies to navigate the mis- and disinformation that increasingly circulates online. Access to online data is inequitable and so too is access to the tools required for sense- and meaning-making. These challenges are addressed here through the lens of event analytics.

Chapter 10 investigates the reasons why news about key events spreads in particular ways, and specifically what barriers might exist to the equitable transmission of information online. It considers three types of events: those that are planned and expected, for example large sporting tournaments; those that are sudden and potentially traumatic, for example natural disasters; and those that are both ongoing and unpredictable, for example climate change and pandemic disease. The research takes into account a range of factors that might influence online news propagation across languages, including regional economic maturity, the role of time zones, geographical variations, and political and cultural differences. Chapter 11 tackles the increasingly important topic of online misinformation, considering not just text or images but adopting a multimodal perspective that reflects the blending of text and image that characterises contemporary social media. Its emphasis is on claim detection, with a view to prioritising data for fact-checking, and notably on the role of images in either underpinning textual claims or functioning as claims in their own right. The dataset of more than 3000 manually annotated tweets collected for the research—in Arabic and English—shares with Chap. 10 a focus on climate change and the COVID-19 pandemic, both highly contested in online spaces, but adds the broader and less obviously polarising subject of technology. Of the three chapters in this section, the final one (Chap. 12) engages most explicitly with storytelling and questions of narrativisation. It is concerned with improving user access to information about events by developing a pipeline to generate narratives from knowledge graphs. The creation of a knowledge graph becomes a starting point for effective and engaging communication about events, in this instance in English and Spanish. Themes of dissemination, accessibility, narrativity, explicability and trust connect the three contributions in this section and suggest promising avenues for future research.

Jane Winters

Chapter 10
Analysis of Event-Centric News Spreading Barriers

Abdul Sittar, Dunja Mladenić, and Marko Grobelnik

Abstract The nature of the topics being discussed in the news is an essential research question in news spreading since it plays a vital role in individual consumer decisions and political and economic interactions. Also of interest is the question of how the news can be spread more widely across multiple barriers, including linguistic, economic, geographical, political, time zone and cultural. Observing event-centric news, we can see that it has different influences on the public and differs in the way that it spreads. For instance, one would expect news regarding natural disasters such as earthquakes to be mostly objective. Climate change, including global warming and pollution, is a very controversial topic, with political interests of different actors at play. Thus, the reporting is expected to be selective and biased. Finally, when sports intersect with larger societal issues, politics can become intertwined with sports news. Analysing the multi-faceted and spatio-temporal aspects of news coverage can bring insights into what may influence the differences in spreading patterns. This chapter will explain the novel analytical methods used to analyse and understand (1) the news-spreading barriers across different cultures and languages, (2) the multi-faceted and spatio-temporal aspects of the news coverage, and (3) news reporting differences across different political alignments and economic conditions.

A. Sittar (✉) · D. Mladenić
Jožef Stefan International Postgraduate School (IPS) and Jozef Stefan Institute, Ljubljana, Slovenia
e-mail: abdul.sittar@ijs.si; dunja.mladenic@ijs.si

M. Grobelnik
Jozef Stefan Institute, Ljubljana, Slovenia
e-mail: marko.grobelnik@ijs.si

I. Marenzi et al. (eds.), *Event Analytics across Languages and Communities*,
https://doi.org/10.1007/978-3-031-64451-1_10

10.1 Introduction

The term "barrier" refers to the fences that are in place between different societies, nations and countries that affect the transfer of information. To put it another way, the storylines of the news are anchored to time, places or entities, and therefore the coverage of news is hampered by many barriers including cultural, economic, political, linguistic or geographical (Sittar et al. 2022b; Rospocher et al. 2016). The roots of the existence of these barriers relate to their influences. The dictionary meaning of the word "influence" is *to cause someone to change a behaviour, belief or opinion*, and in general there are six weapons of influence that advertisers or companies try to use: reciprocity, consistency, social proof, likeability, authority and scarcity. Although compliance practitioners use thousands of different tactics to influence, the majority fall into six basic categories (Muscanell et al. 2014). Each of these categories has the potential to elicit a distinct type of automatic, mindless compliance from people, i.e. a willingness to say yes without first thinking. Here, we will consider the influence of barriers based on two sub-topics, news reporting and news spreading.

There are plenty of factors that influence news reporting and spreading, including media journalists' professional routines (Wu 2007), ownership and control (McChesney and Gasher 2000), political and economic incentives (Hallin and Mancini 2011), legal and regulatory environment (Lowi 1964), technological advancements (Eden 2001) and socio-economic and cultural factors (Galtung 1971). We explain the novel analytical methods used to analyse and understand the influence of the news-spreading barriers. These methods include cascading analysis to compare the cascading chains of different news events; temporal cascading analysis to explore the temporal propagation across different events; quantitative analysis to consider the amount of news propagation across economic, geographic, time zone, political and cultural barriers; and news reporting differences across varying socio-political and economic contexts.

10.2 Methods to Analyse News Spreading

Methods for analysing news spreading are critical for understanding the role that news plays in society and for developing strategies for delivering news that is accurate, informative and impactful. They can also help us gain insights into the patterns and dynamics of how news is shared and disseminated (Bakshy et al. 2012; Sittar et al. 2023). There are several methods and tools available for analysing news spreading, such as network analysis, sentiment analysis and machine learning classifiers. We provide an overview of the tested methods, including information cascading, quantitative analysis and qualitative analysis. We collected the data using the global media monitoring system Event Registry (Leban et al. 2014) and used different types of event-centric news for all these methods. This platform collects similar multilingual news articles from tens of thousands of news sources and identifies events (Leban et al. 2014). It collects data using the News Feed service (Trampuš and Novak 2012), which collects news articles from around 75,000 news sources in various languages (English, German, Spanish and Chinese). To construct an event, it groups similar news articles together. It calculates many features, one of which is the cross-lingual similarity of articles. It does not use any machine translators but rather tries to frame the problem of finding similarities among cross-lingual news articles such that well-established machine learning tools, designed for mono-lingual text-mining tasks, can be used. Regarding the information cascading and quantitative analysis, we utilised global warming, earthquake (Cui et al. 2020) and FIFA World Cup news data (from 2015 to 2020) (Sittar et al. 2020). The three types of events were chosen based on their popularity and diversity. A list of sub-events was observed from top Web sites related to the three events, and we selected those that were most popular in the countries with the selected national languages (Slovene, German, Spanish, Portuguese and English). For the qualitative methods, we utilised news related to the COVID-19 pandemic (Sittar et al. 2022a). The rest of this chapter is dedicated to the presentation of the three chosen methods. More specifically, the first section introduces information cascading and the mechanism for creating cascading chains; the second section provides an overview of the quantitative analysis of news spreading across different barriers; the third section presents qualitative methods to analyse news reporting differences across political and economic barriers; and finally, the fourth section offers conclusions and identifies possible future directions for research.

10.2.1 Cascading Analysis

Information cascading is the process of connecting information sources with the rest of the nodes on social networks (Yang et al. 2020). It unfolds the facts for estimating the importance of an event for a specific language. It further provides a basis for understanding the extent to which an event is important for a country or a region. Cross-lingual information cascading enables us to find the importance and interest for each language group of a specific event. This information chain offers an incomplete picture of the spread; therefore, multiple approaches exist, from a generative approach to probabilistic statistics (Alamsyah and Sonia 2021). The objective of this method is to understand the mechanism of information cascading for event-centric news using the length of cascading chains.

The length of cascading chains refers to the number of steps or links in a chain of information transmission. In the context of information cascades, a chain refers to the sequence of decisions made by individuals as they observe and respond to the actions of others. The length of the chain can have a significant impact on the outcome of the cascade. For example, a longer chain may be prone to errors or biases, as each person in the chain relies on the information and decisions of those before them. On the other hand, a shorter chain may be more efficient

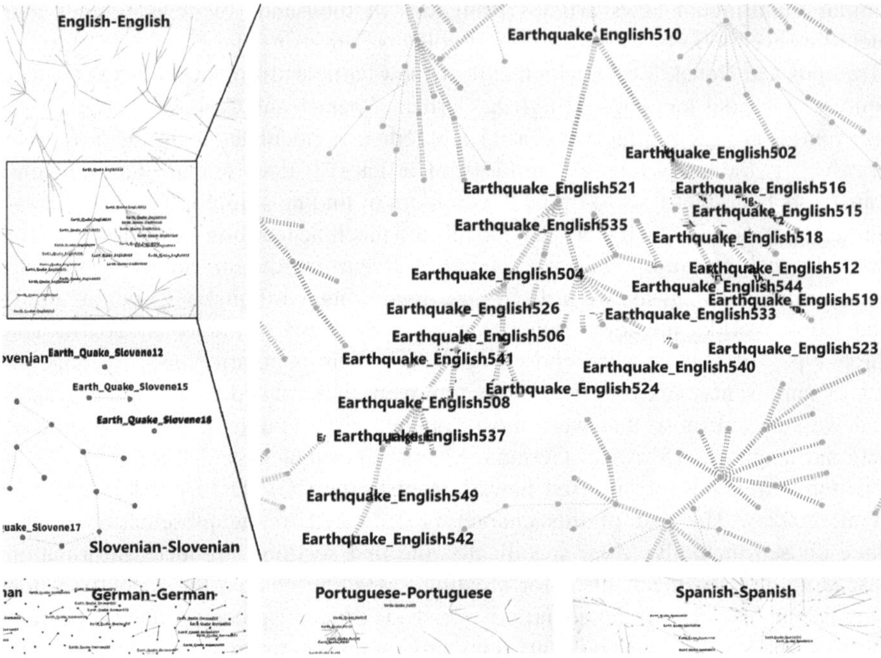

Fig. 10.1 Overview of longest cascading chains in different languages related to events such as cascading chains in English, English and German for the event earthquake

and less vulnerable to errors but may also be less representative of the diverse views and opinions of the group. The length of cascading chains is an important factor to consider when studying and understanding the dynamics of information cascades. Figure 10.1 presents the longest cascading chains of different languages in networks of similar news articles related to the earthquake news event. Figure 10.2 shows the size of communities across different languages. For the earthquake event, Slovene, Portuguese, Spanish, German and English had 0, 15, 31, 101 and 49 chains, respectively (Sittar et al. 2022b).

10.2.1.1 Temporal Cascading Analysis

Temporal cascading analysis is important to understand how quickly event-centric news spreads. Our method presents a new approach to cascading based on news spreading. Firstly, we can observe information flow in mono- and cross-lingual settings. Secondly, we can go beyond information flow based on textual similarity and show the flow of news related to events in different domains and in different languages from the point of view of temporal elements (e.g. monthly spread of

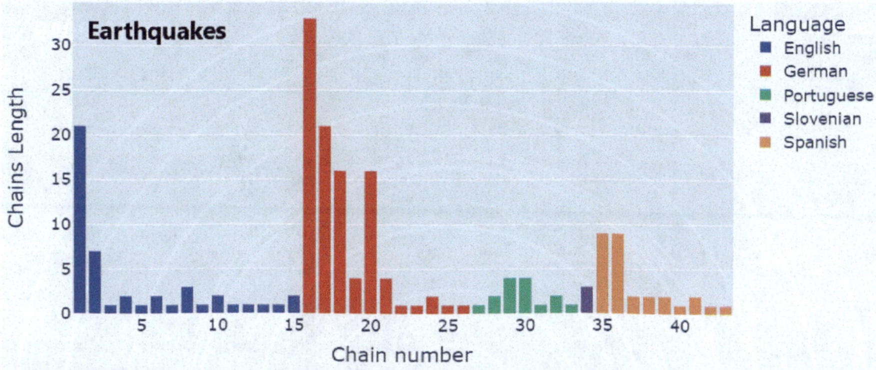

Fig. 10.2 Overview of length of chains for the event earthquake

news) (Brank et al. 2017). Figure 10.3 shows the visual depiction of multilingual temporal propagation for the news event FIFA World Cup. This visualisation takes pairs of news articles as input, draws spirals of articles published within the same month and links the news articles that propagate information from one month to another. There are a total of 72 spirals, which means 6 years, as each spiral represents 1 month. Each dot represents a news article, where colour represents the language of the news article. The connection between the two dots indicates that one article is spreading news to the other article.

10.2.2 Quantitative Analysis

Quantitative analysis can be used to understand the news-spreading behaviour and performance across different languages, economies, time zones, countries, political spectrums and cultures. This analysis can help researchers to understand patterns or trends in the news landscape, identify biases or agendas and measure the effectiveness of news interventions. It can be used in conjunction with qualitative analysis, which involves studying the meaning or context of news through techniques such as content analysis. We present examples of news propagation related to earthquake, FIFA World Cup and global warming across different barriers. The news about these events is collected from the Event Registry global media monitoring system. The information about different barriers is collected from Wikipedia. To identify information propagation, we measure similarity among the news articles using Wikipedia concepts. And we follow the feed-forward mechanism of creating an information propagation network, linking a news article to those published subsequently.

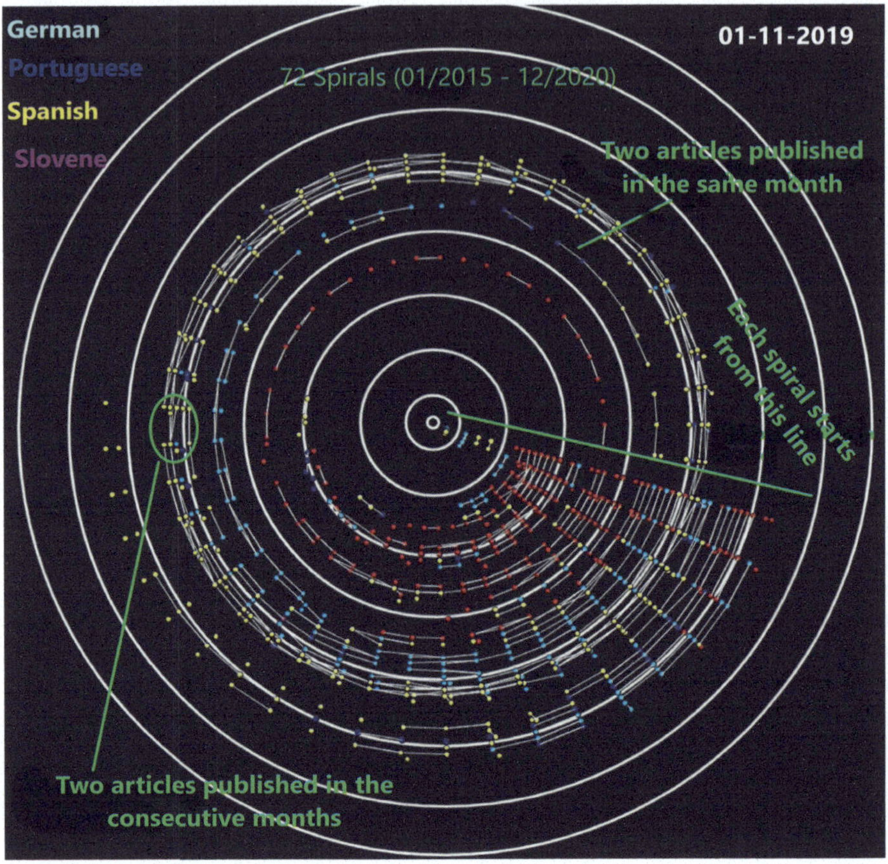

Fig. 10.3 Visual depiction of multilingual temporal propagation for the event FIFA World Cup

10.2.2.1 News Propagation Across Economies

Propagation of news among countries with a certain type of economy can be analysed using different visualisations. Figure 10.4 shows the news propagation related to global warming across different economies using a sankey diagram. As we can see, there are numerous news articles propagating from high-income to high-income countries, whereas little news propagates to lower-middle- and upper-middle-income countries. However, few news articles propagate from lower-middle- and upper-middle-income countries to lower-middle- and upper-middle-income countries.

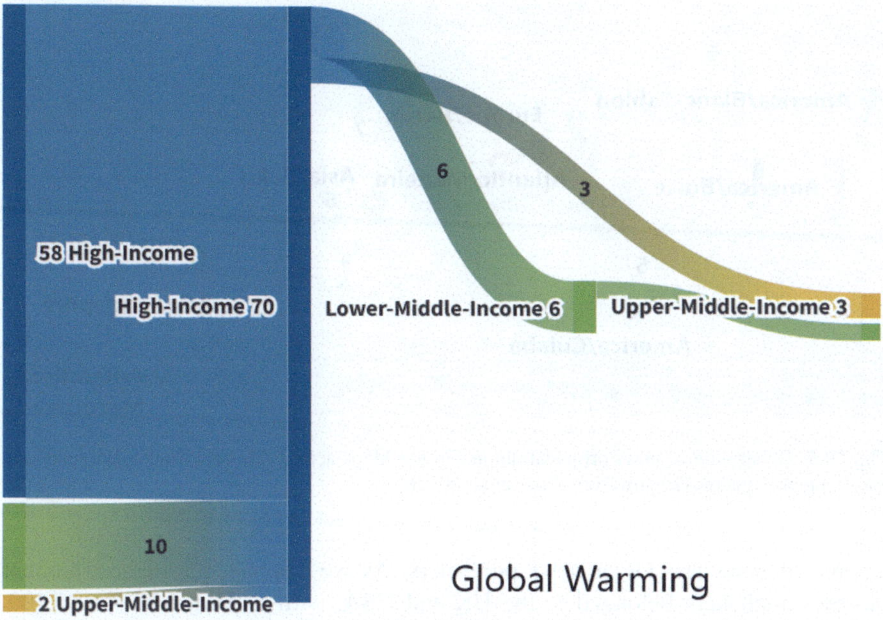

Fig. 10.4 Illustration of news spreading across different income levels (from top blue to bottom yellow: High-Income, Lower-Middle-Income, Upper-Middle-Income and Low-Income) related to the event of global warming

10.2.2.2 News Propagation Across Time Zones

To facilitate the dissemination of earthquake-related news on Google Maps across various time zones, we depict the connections among different time zones by illustrating both the variance in time zone locations (refer to Fig. 10.5) and the Coordinated Universal Time (UTC) (refer to Fig. 10.6) through chord diagrams. The width of each arc represents the number of news articles published from that time zone. The thickness of the connecting ribbons corresponds to the number of news articles propagating between different time zones. As we can see, European countries such as Portugal, the United Kingdom, Germany and Switzerland are surfaced as the most popular in spreading news articles to other countries such as Taiwan, Canada, the United States, Australia and Israel. The difference in time zones among these countries lies between 3–13 hours.

10.2.2.3 News Propagation Across Countries

Google Maps displays the number of news publishers (above) and total number of news articles (below) for the event FIFA World Cup (see Fig. 10.7). The colour of the markers—red, yellow and green—represents the significant, medium and smaller

Fig. 10.5 Illustration of news propagation on Google Maps across different time zones related to the event earthquakes (Figure from Sittar et al. 2022b)

amount of news articles or news publishers. As we can see, the highest number of news publishers belonged to the UK and USA, with a count of 134 and 109, respectively. Countries such as Australia, Canada, Germany, Pakistan, Portugal, Switzerland and the UAE stood in the medium category. Overall, it appears that countries that have a large geographical area have more organisations publishing news about earthquake events than the FIFA World Cup event.

10.2.2.4 News Propagation Across the Political Spectrum

The distribution of political alignments has been presented in Table 10.1. It contains political alignments of news publishers across three different types of events, including the FIFA World Cup, earthquakes and global warming. As we can see, each publisher resides in one of the ten classes displayed in Table 10.1. One significant finding is that the news related to earthquakes was only published by those publishers that were politically neutral, progressive and impartial. Publishers with political alignments designated as having anti-communist, pluralist and new-left political ideas were found publishing more news related to global warming.

10.2.2.5 News Propagation Across Different Cultures

The representation of culture follows the six Hofstede national cultural dimensions, applied across all the countries (Sittar et al. 2022b). These six dimensions are power distance index (PDI), uncertainty avoidance by individuals, non-individualistic cultures, masculinity vs. femininity, long-term orientation and indulgence vs. restraint. The chord diagram visualises the relationship between countries for one cultural

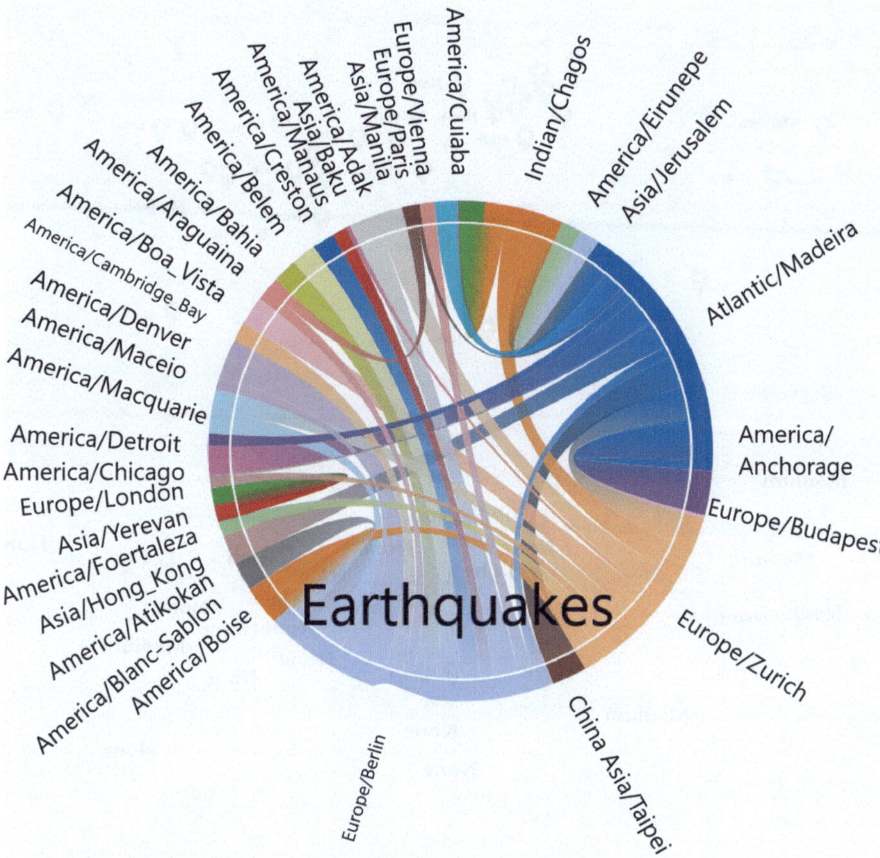

Fig. 10.6 Propagation depiction across different time zones illustrated with both a difference among time zone locations and UTC time

dimension: the power distance index for the event of global warming (see Fig. 10.8). There are four categories of this dimension: low, upper lower, upper higher and high. As we can see, the countries in the low category only propagate news to those countries that stand in the upper-lower and upper-higher categories (indicated with blue lines), whereas countries in the high category only propagate news to those in the upper-higher category (indicated with orange lines).

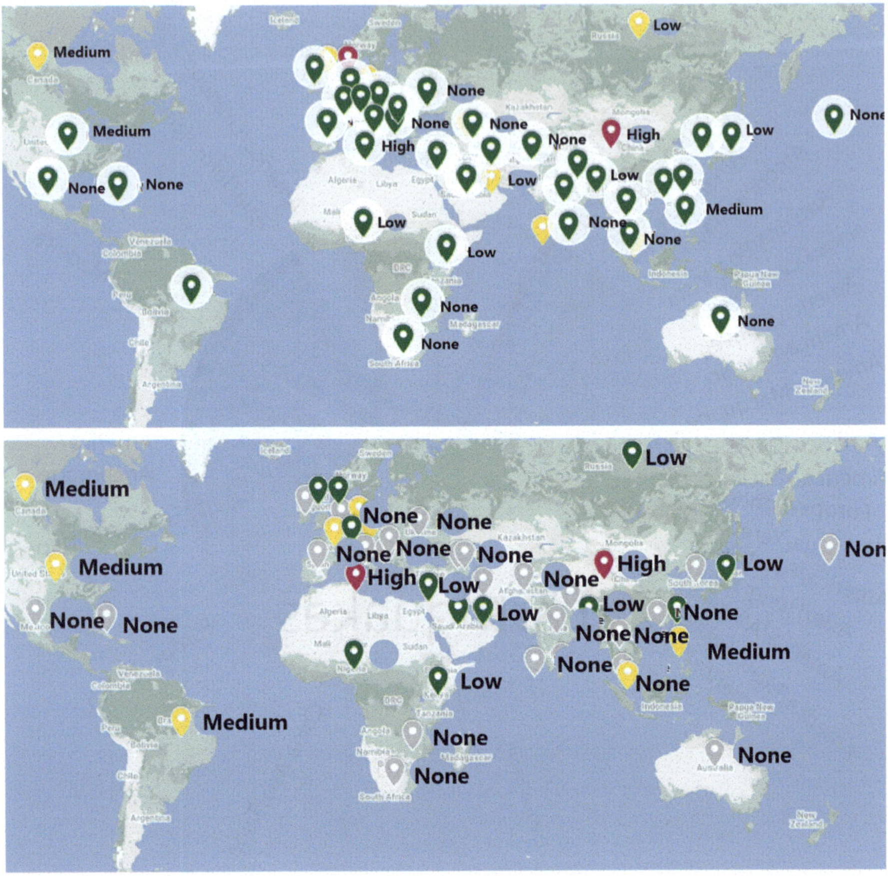

Fig. 10.7 An illustration of publishers' (first row) and articles' (second row) distribution over Google Maps for the event FIFA World Cup. Red, yellow and green colours show the significant, medium and smaller amounts of articles or news publishers (Figure from Sittar et al. 2022b)

10.2.3 *Qualitative Analysis*

Qualitative analysis can be used to understand the context of news-reporting differences. These differences refer to the ways in which news is produced and disseminated across different countries, cultures or political alignments. These differences may be due to a variety of factors, such as the level of media freedom, the availability of technology, cultural values and beliefs or political agendas (Sittar and Mladenic 2023). News-reporting differences can result in variations in the types of news that are covered, the way that news is presented, the sources or perspectives that are included or the tone or language used (Sittar et al. 2022c). These differences can have significant impacts on the way that news is received and interpreted by

Table 10.1 Proclivity of news publishers with a specific political class toward different events

No.	Event type	Classes of political alignment
1	Global warming	Anti-communist
2	Global warming, Earthquake	Catholic
3	Global warming, FIFA World Cup, earthquake	Centrism
4	FIFA World Cup, global warming	Conservative
5	FIFA World Cup, global warming	Independent
6	FIFA World Cup, global warming, earthquake	Liberalism
7	Global warming	New Left
8	Global warming	Pluralism
9	FIFA World Cup, global warming, earthquake	Social Liberalism
10	FIFA World Cup, global warming	Left Wing
11	FIFA World Cup	Centre Right
12	FIFA World Cup	Moderate
13	FIFA World Cup	Progressive
14	Earthquake	Impartiality
15	Earthquake	Progressive
16	Earthquake	Neutral

different audiences and can shape public opinion and influence political or cultural outcomes.

By understanding the differences in news reporting, journalists can improve the accuracy and fairness of their reporting and can ensure that they are providing a balanced and nuanced portrayal of global events and issues. This can also help enhance media literacy and critical thinking skills, which can enable people to better evaluate the credibility and reliability of different news sources. Policymaking can gain a more complete and accurate understanding of global events and issues and can make more informed decisions. Lastly, greater insight can be helpful in marketing and advertising to tailor messages and campaigns to different cultural audiences and for better understanding the cultural influences on news consumption.

10.2.3.1 News Reporting Differences Across Different Political Alignments and Economies

Spatio-temporal analysis is important in understanding the political and economic barriers that influence the production and dissemination of news because it enables researchers to examine the intersection of space (geography) and time (history) in the flow of information. By analysing the spatio-temporal dynamics of news propagation, researchers can identify patterns and trends in the way that news spreads and the factors that influence its spread, such as geography, culture, economy or political alignment (Camaj 2010). This can help researchers understand how political and economic barriers, such as censorship or media ownership, shape

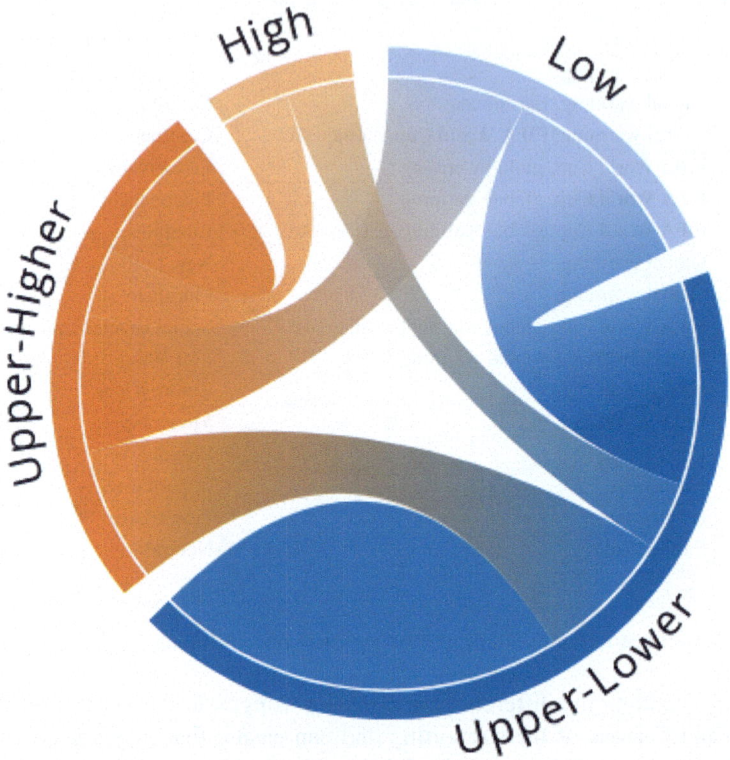

Fig. 10.8 News propagation visualisation for one cultural dimension (PDI) for the event global warming

the news landscape and the flow of information. Spatio-temporal analysis can also help researchers identify opportunities to improve the accuracy and timeliness of news and to reduce biases or agendas in the news.

To understand news-reporting differences across different political alignments and economies, the first key criterion is the availability of sufficient trustworthy data about an event. Second, it is important to know the political alignments of different news publishers, which is not publicly available except for a few of the most read news publishers, i.e. Daily Mail, Dawn and USA Today. In order to understand the economic condition of the location of news publishers, it is best to have local economic representation, but this has also not been measured and made available for most of the countries across the globe (Sittar and Mladenic 2021). Finally, the results of these differences can be a list of topics or latent themes from a large amount of news text. There are multiple topic modelling techniques available to analyse these.

A dataset of news articles (24,000) was collected from the global media monitoring platform Event Registry about the COVID-19 pandemic (from January 2020 to May 2021). Since COVID-19 had major effects on different economies and was associated with many conspiracies and fake information, we take into account only two factors, political and economic. Analysing the frequent topics across different political alignments can help us find correlations between topics and political alignments. Therefore, it might be possible to see whether or not political alignment is associated with a particular effect on news spreading related to COVID-19. Similarly, analysing the frequent topics across different levels of economic prosperity can help us see the correlations between certain topics and different levels of economic prosperity (Büyüksarıkulak and Kahramanoğlu 2019). Therefore, it might be possible to see if a country's economic situation has a particular effect on news spreading related to COVID-19. This dataset is based on the top-ten newspapers (dailymail.co.uk, dawn.com, irishtimes.com, livemint.com, mainichi.jp, theglobeandmail.com, thegaurdian.com, thejakartapost.com, usatoday.com, yenisafak.com) belonging to different political alignments and from different economic backgrounds (Sittar et al. 2022a). These newspapers were selected based on the following pre-conditions: (1) at least a few news articles had to published by a newspaper for each month (January 2020 to May 2021), and (2) the newspapers had to belong to countries from different economic backgrounds.

To infer a list of topics or latent themes from a corpus, we proposed enhanced topic modelling techniques with the pooling of news articles based on user queries. The results are more coherent if the text is semantically similar. Therefore, the motivation behind our proposed method is to pool the news articles based on user queries in order to extract the most relevant latent themes without modifying the basic structure of LDA. Previous studies used pooling based on other parameters. For example, pooling has been applied to Twitter datasets based on hashtags (Mehrotra et al. 2013), while (Alvarez-Melis and Saveski 2016) identified relevant conversations and applied pooling on tweets. However, the problem of pooling based on user queries is not explored for news articles. To fill this gap, the present study aims to identify the coherence score differences with and without pooling based on user queries (Sittar et al. 2022a). The popular user queries were as follows: (1) lab leak theory; (2) efficacy of vaccines; (3) lockdown policies and efficiency; (4) seriousness of COVID-19 virus; and (5) if masks can protect against COVID.

Figure 10.9 shows the word clouds of frequent topics for different political alignments, and Fig. 10.10 shows the word clouds of frequent topics for different economic levels. These results are based on an enhanced topic-modelling technique, with the pooling of news articles based on the above-mentioned user queries. The overall findings on economic issues, which appear as a consequence of COVID-19, are quite consistent with the economic situation of the country of each newspaper, whereas the political alignment of a newspaper does not show a clear consistency in relation to the present topic. However, it also appears that political alignment influences news reporting and newspapers report on national economic situations more than global economic situations.

Similarly, we applied the enhanced topic-modelling approach to the Russia-Ukraine war to confirm whether the proposed methodology is suitable only for COVID-19 or whether it can be applied to different controversial events. Similarly

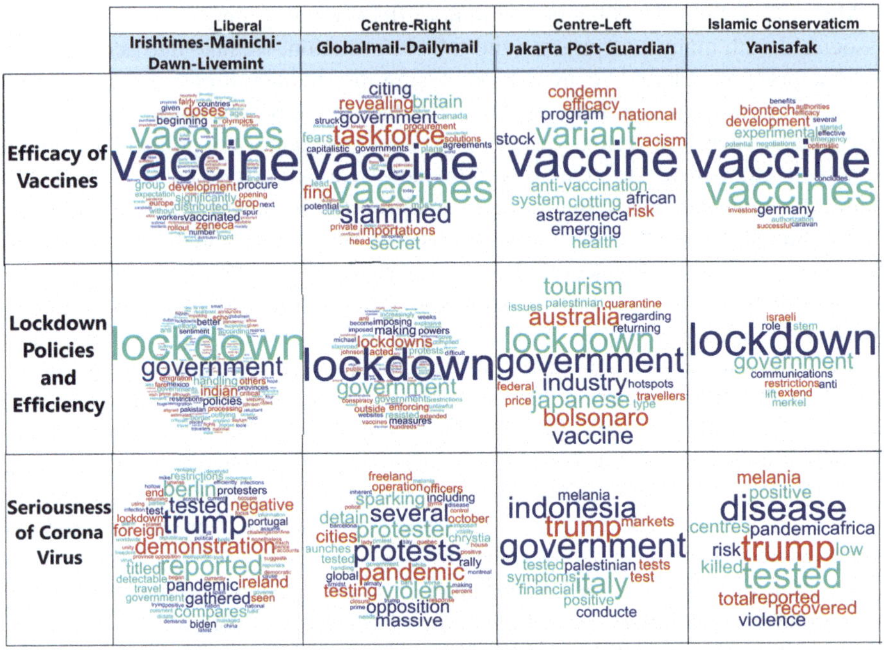

Fig. 10.9 Word clouds showing the keywords appearing most frequently related to five queries in relation to different political alignments

to COVID-19, our objective was to understand the effect of different political and economic contexts on news reporting. We collected the dataset of news articles that were published by the same newspapers as for COVID-19 (dailymail.co.uk, dawn.com, irishtimes.com, livemint.com, mainichi.jp, theglobeandmail.com, thejakartapost.com, usatoday.com, yenisafak.com), but we selected a short timeline of two months (January and February 2023) for this event. Because of this short timeline, Event Registry did not have enough articles from The Guardian newspaper for it to be considered for comparison, and consequently it was excluded. The collected dataset for this event consists of 1512 news articles. We identified two main queries to pool news articles for the two months: (1) What is the refugee crisis in Ukraine? (2) What are the implications of the Russian-Ukraine war?

Figure 10.11 shows the word clouds for different political alignments along with all the queries. Regarding the refugee crisis in Ukraine, liberal newspapers have been found discussing the global crisis, economic and political uncertainty, the energy crisis, stock market rates, inflation, wheat prices and food shortages. Right-wing newspapers appeared to focus more on sanctions, peace-talk proposals, oil prices and the fuel market, whereas left-wing newspapers talk about global prices. Regarding the implications of the Russian-Ukraine war, the newspapers with a liberal political alignment mention the topics referendum, cutting oil production, military force, nuclear agreement, global-inflation, price growth, global crisis,

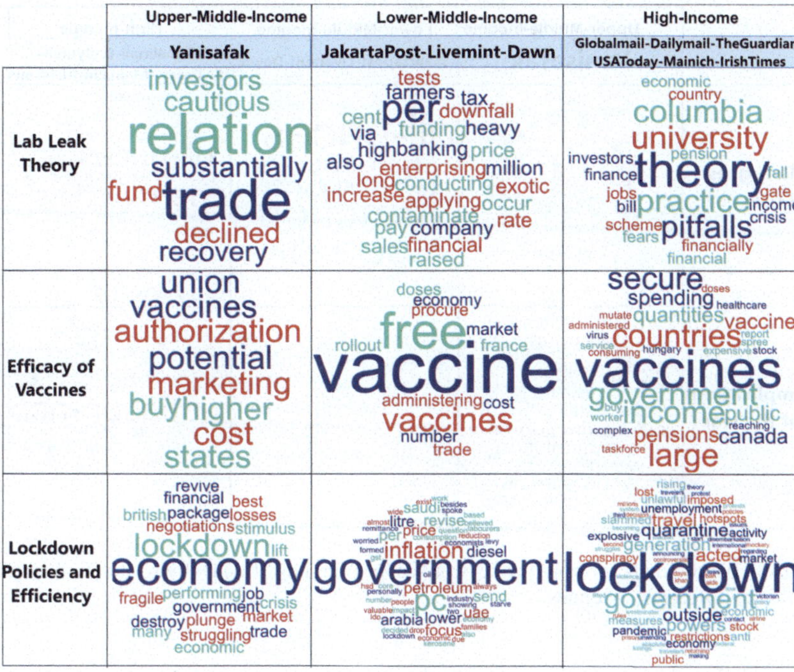

Fig. 10.10 Word clouds showing the keywords appearing most frequently related to five queries in relation to different economies

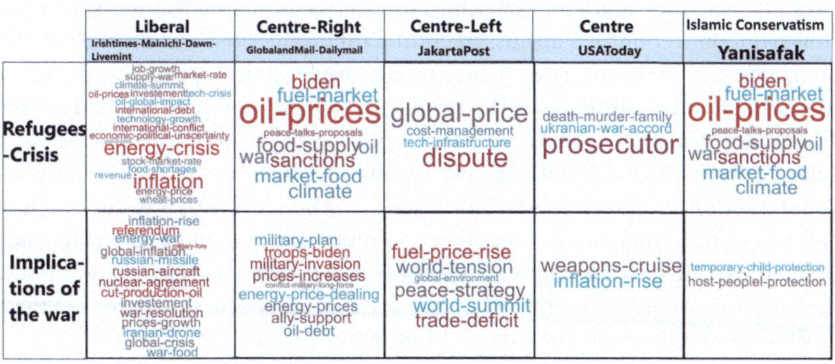

Fig. 10.11 Word clouds showing the keywords appearing most frequently related to five queries in relation to different political alignments

inflation rise and energy war. Figure 10.12 shows the word clouds for different economic levels along with all the queries. Regarding the refugee crisis in Ukraine, the topics mentioned in low-income countries were international debt, economic uncertainty, market rates, investment, global crisis, inflation, wheat prices, revenue, global prices and cost management. Topics such as oil prices, sea oil, sanctions

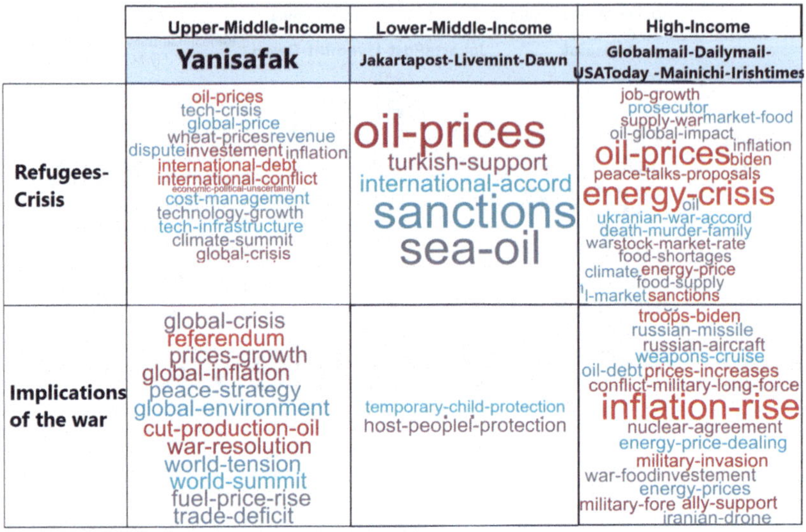

Fig. 10.12 Word clouds showing the keywords appearing most frequently related to five queries in relation to different economies

and international accord appeared with high frequency in middle-income countries. Topics that appeared in newspapers from high-income countries are sanctions, peace talks, energy crisis, stock market rate, inflation, food shortages, global impact of oil, food supply and fuel market. Regarding the implications of the Russian-Ukraine war, lower-income countries discussed cutting oil production, war resolution, global inflation, price growth, the rise in fuel prices, trade deficit, peace strategy, world tensions and global environment. The results across different political and economic conditions are not quite consistent in the case of the Ukraine and Russian war, whereas in the case of COVID-19, the results are more consistent. Firstly, event type matters a lot in news reporting differences. The COVID-19 event is a natural disaster where reporting can be smoother and more transparent, whereas for the war event, the reporting may be political. Secondly, the timeline is shorter for the second event, consisting of only 2 months (January and February 2023), whereas in the case of COVID-19, the timeline consists of 14 months (January 2020 to May 2021).

10.3 Conclusions and Future Work

The results discussed in this chapter show the potential impact of the proposed topic modelling technique and the analytical methods used for the analysis of news-spreading barriers. Overall, our findings suggest that news published in news articles propagates to a greater degree across languages for natural disasters (earthquake events) than climate change (global warming) and sports (FIFA World Cup). In our experiments, we observed more cascading chains as well as longer cascading chains for earthquakes. The results of quantitative analysis suggest that for global warming events, more news has propagated among economically strong countries than to economically weaker countries. The results of qualitative analysis (news reporting differences) across political and economic barriers suggest that the economic issues that appear as a consequence of COVID-19 are quite consistent with the economic situation of the country of the newspaper.

We are now working on developing a system to automatically classify the barriers across different types of events (business, computers, games, health, home, recreation, science, shopping, society and sports). The annotation for each barrier will be based on news metadata and semantic similarity. Moreover, we are investigating suitable features for this classification including Wikipedia concepts, the sentiment of the news, news headlines and common sense knowledge.

Acknowledgments The work presented in this chapter has received funding from the European Union's Horizon 2020 research and innovation program under the Marie Skłodowska-Curie grant agreement no. 812997 and under the name CLEOPATRA (Cross-lingual Event-centric Open Analytics Research Academy).

References

Alamsyah A, Sonia A (2021) Information cascade mechanism and measurement of Indonesian fake news. In: 2021 9th international conference on information and communication technology (ICoICT). IEEE, pp 566–570

Alvarez-Melis D, Saveski M (2016) Topic modeling in Twitter: aggregating tweets by conversations. In: Proceedings of the international AAAI conference on web and social media, vol 10, pp 519–522

Bakshy E, Rosenn I, Marlow C, Adamic L (2012) The role of social networks in information diffusion. In: Proceedings of the 21st international conference on World Wide Web, pp 519–528

Brank J, Leban G, Grobelnik M (2017) Annotating documents with relevant Wikipedia concepts. Proc SiKDD 472:159–162

Büyüksarıkulak AM, Kahramanoğlu A (2019) The prosperity index and its relationship with economic growth: case of turkey. J Entrepreneurship Bus Econ 7(2):1–30

Camaj L (2010) Media framing through stages of a political discourse: international news agencies' coverage of Kosovo's status negotiations. Int Commun Gaz 72(7):635–653

Cui Y, Ni S, Shen S, Wang Z (2020) Modeling the dynamics of information dissemination under disaster. Phys A: Stat Mech Appl 537:122822

Eden B (2001) The rise of the network society. The information age: economy, society and culture. Bottom Line 14(3):124–162

Galtung J (1971) A structural theory of imperialism. J Peace Res 8(2):81–117

Hallin DC, Mancini P (2011) Comparing media systems beyond the Western world. Cambridge University Press, Cambridge. https://doi.org/10.1017/CBO9781139005098

Leban G, Fortuna B, Brank J, Grobelnik M (2014) Event registry: learning about world events from news. In: Proceedings of the 23rd international conference on World Wide Web, pp 107–110

Lowi TJ (1964) American business, public policy, case-studies, and political theory. World Polit 16(4):677–715

McChesney RW, Gasher M (2000) Rich media, poor democracy: communication politics in dubious times. Can J Commun 25(4):585

Mehrotra R, Sanner S, Buntine W, Xie L (2013) Improving LDA topic models for microblogs via tweet pooling and automatic labeling. In: Proceedings of the 36th international ACM SIGIR conference on research and development in information retrieval, pp 889–892

Muscanell NL, Guadagno RE, Murphy S (2014) Weapons of influence misused: a social influence analysis of why people fall prey to internet scams. Social Personality Psychol. Compass 8(7):388–396

Rospocher M, Van Erp M, Vossen P, Fokkens A, Aldabe I, Rigau G, Soroa A, Ploeger T, Bogaard T (2016) Building event-centric knowledge graphs from news. J Web Semant 37:132–151

Sittar A, Mladenic D (2021) How are the economic conditions and political alignment of a newspaper reflected in the events they report on? In: Central European conference on information and intelligent systems, faculty of organization and informatics Varazdin, pp 201–208

Sittar A, Mladenic D (2023) Classification of cross-cultural news events. Preprint. arXiv:230105543

Sittar A, Mladenic D, Erjavec T (2020) A dataset for information spreading over the news. In: Proceedings of the 23th international multiconference information society SiKDD, vol 100, pp 5–8

Sittar A, Major D, Mello C, Mladenić D, Grobelnik M (2022a) Political and economic patterns in COVID-19 news: from lockdown to vaccination. IEEE Access 10:40036–40050. https://doi.org/10.1109/ACCESS.2022.3164692, https://ieeexplore.ieee.org/abstract/document/9749092

Sittar A, Mladenić D, Grobelnik M (2022b) Analysis of information cascading and propagation barriers across distinctive news events. J Intell Inf Syst 58(1):119–152. https://doi.org/10.1007/s10844-021-00654-9, https://link.springer.com/article/10.1007/s10844-021-00654-9

Sittar A, Webber J, Mladenic D (2022c) Stylistic features in clustering news reporting: news articles on brexit. In: Proceedings of the 23th international multiconference information society SiKDD, vol 100, pp 21–25

Sittar A, Mladenić D, Grobelnik M (2023) Profiling the barriers to the spreading of news using news headlines. Front Artif Intell 6:1225213. https://doi.org/10.3389/frai.2023.1225213, https://www.frontiersin.org/articles/10.3389/frai.2023.1225213/full

Trampuš M, Novak B (2012) Internals of an aggregated web news feed. In: Proceedings of 15th multiconference on information society, pp 221–224

Wu HD (2007) A brave new world for international news? Exploring the determinants of the coverage of foreign news on us websites. Int Commu Gaz 69(6):539–551

Yang G, Csikász-Nagy A, Waites W, Xiao G, Cavaliere M (2020) Information cascades and the collapse of cooperation. Sci Rep 10(1):1–13

Chapter 11
Claim Detection in Social Media

Gullal S. Cheema, Eric Müller-Budack, Christian Otto, and Ralph Ewerth

Abstract In recent years, the problem of misinformation on the web has become widespread across languages, countries and various social media platforms. One problem central to stopping the spread of misinformation is identifying claims and prioritising them for fact-checking. Although there has been much work on automated claim detection from text recently, the role of images and their variety still need to be explored. As posts and content shared on social media are often multimodal, it has become crucial to view the problem of misinformation and fake news from a multimodal perspective. In this chapter, first, we present an overview of existing claim detection methods and their limitations; second, we present a unimodal approach to identify check-worthy claims; third, and lastly, we introduce a dataset that takes both the image and text into account for detecting claims and benchmark recent multimodal models on the task.

11.1 Introduction

In today's digital age, misinformation has become a pervasive problem that proliferates at an alarming rate through various digital platforms, particularly social

G. S. Cheema (✉)
L3S Research Center, Leibniz Universität Hannover, Hannover, Germany
e-mail: gullal.cheema@tib.eu

E. Müller-Budack
TIB—Leibniz Information Centre for Science and Technology, Hannover, Germany
e-mail: eric.mueller@tib.eu

C. Otto
Institute of Medical Statistics, Computer and Data Sciences, Jena University Hospital, Jena, Germany
e-mail: christian.otto@med.uni-jena.de

R. Ewerth
TIB—Leibniz Information Centre for Science and Technology, Hannover, Germany

L3S Research Center, Leibniz Universität Hannover, Hannover, Germany
e-mail: ralph.ewerth@tib.eu

I. Marenzi et al. (eds.), *Event Analytics across Languages and Communities*,
https://doi.org/10.1007/978-3-031-64451-1_11

media. This issue has become even more pressing with recent events like COVID-19 and the Russia-Ukraine war, as false or misleading information can influence public opinion and put people's safety at risk. However, tackling misinformation online and specifically on social media platforms is challenging due to the variety of information and the volume and speed of streaming data. As a consequence, automatic and semi-automatic detection of misinformation and fake news has attracted much interest from the research community in the last decade. Several studies have proposed datasets in multiple languages (D'Ulizia et al. 2021), leveraged news content (Albahar 2021) and social context (Shu et al. 2019; Nguyen et al. 2022), and explored multimodality (Cao et al. 2020; Giachanou et al. 2020a) to tackle the ever-increasing problem of fake news on digital media.

In this chapter, we focus on identifying claims on social media and the significance of images in either supporting claim text or acting as claims themselves. As social media is inherently multimodal in nature, fact-checking initiatives and computation methods consider not only text but also image content (Giachanou et al. 2019; Khattar et al. 2019; Singhal et al. 2019; Wang et al. 2018) as it can be easily fabricated and manipulated due to the availability of free image and video editing tools. Traditionally, claim detection is studied from a linguistic standpoint where both syntax (Rosenthal and McKeown 2012) and semantics (Levy et al. 2014) of the language matter to detect a claim accurately. However, claims or fake news on social media are not bound to just one modality and become a complex problem with additional modalities like images and videos. While it is clear that a claim in the text is denoted in verbal form, it can also be part of the visual content or appear as overlaid text in the image. Even though much effort has been spent on the curation of datasets (Boididou et al. 2016; Nakamura et al. 2020; Jindal et al. 2020) and the development of computational models for multimodal fake news detection on social media (Ajao et al. 2018; Wang et al. 2018; Khattar et al. 2019; Singhal et al. 2019), more efforts are needed to detect and understand claims across multiple modalities (Zlatkova et al. 2019; Cheema et al. 2020b).

For instance, Fig. 11.1 shows examples of tweets that are not claims vs claims that should be prioritised for fact-checking (known as check-worthy (Barrón-Cedeño et al. 2020)). Check-worthy tweets and claims typically involve prominent figures, organisations or nations, as well as specific events and details that could cause controversial discussions or opinions if they are inaccurate. Conversely, tweets that are claims but not check-worthy are typically informative statements or definitions of terms that are unlikely to cause harm if incorrect. On the right, we provide examples of image-text tweet pairs, where it is necessary to identify the claim in the text and extract information from the corresponding image that could potentially make the pair a claim or a check-worthy claim. The two rightmost examples show images with varying amounts of detail. Both have overlaid text, with the top example depicting a scenario (building a hospital) and the bottom example featuring an infographic (a geographical map) that could be manipulated and inaccurate compared to its source.

This chapter presents approaches to identify claims, detect check-worthiness and investigate the role of images in the context of claim detection on social media. In Sect. 11.2, we first provide a detailed overview of recent work and progress in

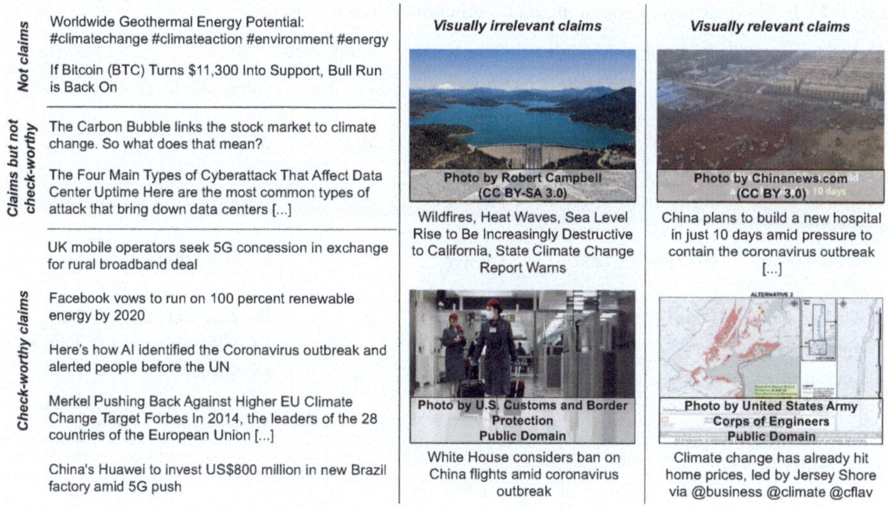

Fig. 11.1 Examples of statements that are not claims, claims and check-worthy claims, and image-text tweet pairs of visually irrelevant and visually relevant claims **Due to possible licensing issues, best alternative images are shown here*

the area of claim detection and multimodal misinformation detection in general. We then present a hybrid approach in Sect. 11.3 that combines lexical features and contextual neural network embeddings to identify check-worthiness in both English and Arabic. In Sect. 11.4, we present a new dataset for multimodal claim detection first introduced in Cheema et al. (2022) that is annotated based on both image and text. Section 11.5 concludes with the main findings, connections to other chapters in the book and outlines areas of future work.

11.2 Background and Related Work

This section provides a review of recent developments in the area of claim detection, starting with text-based claim detection. Following this, we briefly discuss related approaches that consider the role of images in multimodal fake news detection. Finally, we conclude with a discussion of recent approaches that investigate images and their role in claim detection.

11.2.1 Text-Based Approaches

Before research on claim detection targeted social media, pioneering work by Rosenthal and McKeown (2012) focused on claims in *Wikipedia* discussion forums. They used lexical and syntactic features in addition to sentiment and other statistical

Table 11.1 Comparison of social media based claim datasets

Datasets	#Samples	Modality	Data source	Language	Topic	Task(s)
Zlatkova et al. (2019)[a]	1233	Image, Text	Snopes, Reuters	en	Multi-topic	T vs F
Nakov et al. (2021)	18,014[b]	Text	Twitter	m+	Multi-topic[b]	C.W.E
Gupta et al. (2021)	9981	Text	Twitter	en	COVID-19	C.D
Iskender et al. (2021)	300 pairs	Text	Twitter	de	Climate change	C.D, E.D
Wührl and Klinger (2021)	1200	Text	Twitter	en	Biomedical & COVID-19	C.D, C.T.D
Cheema et al. (2022) (Ours)	3400	Image, Text	Twitter	en	COVID-19, Climate change, Technology	C.D, C.W.E, V.R

Best results in bold

[a] Zlatkova et al. (2019) is a mix of actual news photographs (from Reuters) and possibly fake images (from Snopes), which went viral on social media sites like Reddit. m+ stands for multilingual

[b] 1312 samples are in English and only on the topic of COVID-19

Task(s) are represented as: T vs F (True vs False), C.W.E (check-worthiness estimation), C.D (claim detection), E.D (evidence detection), C.T.D (claim type detection), V.R (visual relevance)

features over text. Since then, researchers have proposed context-dependent (Levy et al. 2014), context-independent (Lippi and Torroni 2015), cross-domain (Daxenberger et al. 2017) and in-domain approaches for claim detection. Recently, transformer-based models (Chakrabarty et al. 2019) have replaced structure-based claim detection approaches due to their success in several natural language processing (NLP) downstream tasks. A series of workshops (Barrón-Cedeño et al. 2020; Nakov et al. 2021, 2022) focused on claim detection and verification on Twitter and organised challenges with several sub-tasks on text-based claim detection around the topic of *COVID-19* and several other topics and events in multiple languages. The datasets in these challenges are labelled to predict the most check-worthy claims that should be prioritised for fact-checking to stop the spread of misinformation.

Gupta et al. (2021) addressed the limitations of current methods in cross-domain claim detection by proposing a new dataset of ~10,000 claims on *COVID-19*. They also proposed a model that combines transformer features with learnable syntactic feature embeddings. Another dataset introduced by Iskender et al. (2021) includes tweets in German about *climate change* for claim and evidence detection. Wührl and Klinger (2021) created a dataset for biomedical Twitter claims related to *COVID-19, measles, cystic fibrosis* and *depression*. One common theme and challenge among all the datasets is the variety of claims, where some types of claims (like implicit) are harder to detect than explicit ones where a typical claim structure is present. Table 11.1 shows a comparison of existing social-media-based claim datasets, with number of samples, modalities, data sources, language, topic and type of tasks.

11.2.2 Multimodal Approaches

From the multimodal perspective, very few works have analysed the role of images in the context of claims. Zlatkova et al. (2019) introduced a dataset that consists of claims and is created from the idea of investigating questionable or outright false images which supplement fake news or claims. The authors used reverse image search and several image metadata features such as tags from the Google Vision API, URL domains and categories, reliability of the image source, etc. Similarly, Wang et al. (2020) performed a large-scale study by analysing manipulated or misleading images in news discussions on forums like *Reddit*, *4chan* and *Twitter*. For claim detection, Cheema et al. (2021) extended the text-based claim detection datasets of Barrón-Cedeño et al. (2020) and Gupta et al. (2021) with images to evaluate multimodal detection approaches. Although previous work has provided multimodal datasets on claims, they are either on the veracity (true or false) of claims or labelled only text-based for a single topic (COVID-19).

For multimodal fake news in general, several benchmark datasets have been proposed in the last few years, generating interest in developing multimodal visual and textual models. In one of the relatively early works, Jin et al. (2017) explored rumour detection on Twitter using text, social context (emoticons, URLs, hashtags) and the image by learning a joint representation with attention from LSTM outputs over image features. The authors observed the benefit of using the image and social context in addition to text by improving the detection of fake news in Twitter and Weibo datasets. Later, Wang et al. (2018) proposed an improved model that learns a multi-task model to detect fake news as one task and event discriminator as another task to learn event invariant representations. Since then, improvements have been proposed via using multimodal variational autoencoders (Khattar et al. 2019), transfer learning (Giachanou et al. 2020b; Singhal et al. 2019) with transformer-based text and deep visual CNN models. Recently, Nakamura et al. (2020) proposed a fake news dataset, *r/Fakeddit* mined from Reddit, with over 1 million samples, which includes text, images, metadata and comments data. The data is labelled through distant supervision into 2-way, 3-way and 6-way classification categories.

11.3 Text-Based Claim Detection

In this section, we introduce and summarise a text-based approach (Cheema et al. 2020a) for claim check-worthiness estimation. Check-worthiness estimation is the task of predicting whether a tweet includes a claim that is of interest to a large audience. We present a solution that works for both English [43] and Arabic [19] tweets. Our approach (see Fig. 11.2) is motivated by the successful use of lexical, syntactic and contextual features in the previous editions of the CheckThat! check-worthiness task for political debates. We explore the fusion of syntactic features and deep transformer Bidirectional Encoder Representations from Transformers (BERT) embeddings to classify the check-worthiness of a tweet. We use part-of-speech (POS) tags, named entities and dependency relations as syntactic features and a combination of hidden layers in BERT to compute tweet embedding.

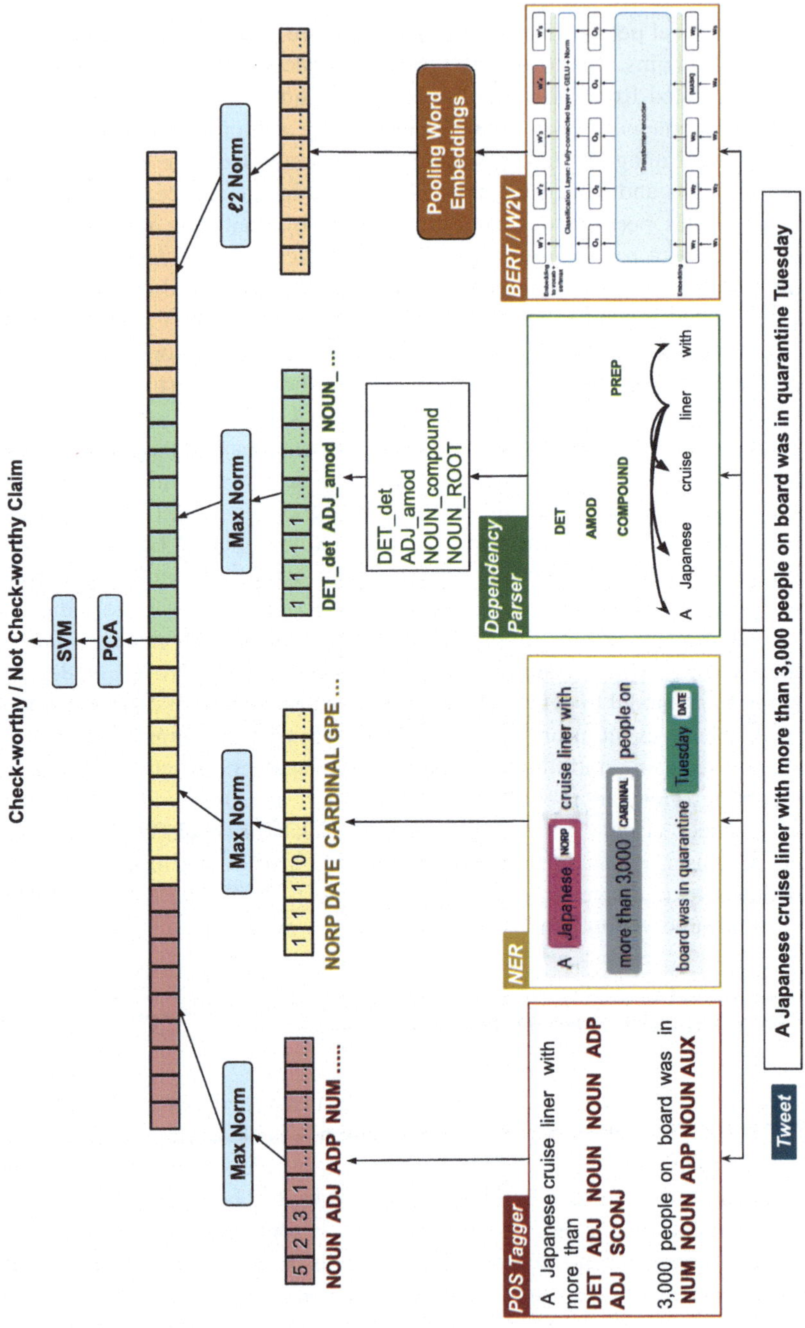

Fig. 11.2 Approach for text-based claim check-worthiness estimation

Before learning the model with a Support Vector Machine (SVM) (Suykens and Vandewalle 1999), we use Principal Component Analysis (PCA) (Wold et al. 1987) for dimensionality reduction.

The approach is benchmarked on two datasets from the CheckThat! 2020 challenge (Barrón-Cedeño et al. 2020). Given that this task contains less training data, we approach this problem with the idea of creating a rich feature representation, reducing the dimensions of a large feature set with PCA (Wold et al. 1987) and then learning the model with an SVM. In doing so, our goal is also to understand which features are the most important for check-worthiness estimation from tweet content. As context is very important for downstream NLP tasks, we experiment with word embeddings (word2vec (Mikolov et al. 2013), GloVe (Pennington et al. 2014)) and BERT (Devlin et al. 2019) embeddings to create a sentence representation of each tweet. Our pre-processing and feature extraction is agnostic to the topic of the tweet so that it can be applied to any domain. Next, we summarise the feature extraction and experimental results, which are discussed in more detail in Cheema et al. (2020a).

11.3.1 Feature Extraction

This subsection describes the type of syntactic features and contextual embeddings extracted from a transformer model used in our proposed approach.

11.3.1.1 Syntactic Features

We use the following syntactic features for English and Arabic tasks: *parts-of-speech (POS) tags*, *named entities (NE)* and *dependency parse tree relations*. For both English and Arabic, we extract 16 **POS tags** in total and through our empirical evaluation we find the following eight tags to be the most useful when used as features: NOUN, VERB, PROPN, ADJ, ADV, NUM, ADP, PRON. For Arabic, the additional four tags are useful features: DET, INTJ, AUX, PART. We used the chosen set of POS tags for each respective language to encode the syntactic information of tweets. For **named entities**, we evaluated and found the following named entity types to be the most important features: (GPE, PERSON, ORG, NORP, LOC, DATE, CARDINAL, TIME, ORDINAL, FAC, MONEY) for English and (LOC, PER, ORG, MISC) for Arabic. We also found that while developing feature combinations named entities do not add much value to overall accuracy, and hence our best performing model does not include them. Lastly, **syntactic dependency** features are constructed using the dependency relation between tokens in a given tweet. We use the dependency relation between two nodes in the parsed tree if the child and parent nodes' POS tags are one of the following: ADJ, ADV, NOUN, PROPN, VERB or NUM. All dependency relations that match the defined constraint are converted into the triplet relation, such as (*child node-POS, dependency-relation, parent-POS*), and pairs, such as (*child node-POS, dependency-relation*), where the relation is not part of a feature representation. We found that the features based on pairs of child and parent node perform better than the triplet feature. For encoding a feature, we get a histogram vector which contains the number of types of POS tag, named entity or syntactic relation pair. Finally, we normalise each type of feature with maximum value (Max Norm) in the vector. The process of feature encoding is also shown in Fig. 11.2.

11.3.1.2 Contextual Features

One simple way to get a contextual representation of a sentence is to average the word embeddings of each token in a given sentence. For this purpose, we experiment with three types of word embeddings pre-trained on three different sources for English: GloVe embeddings (Pennington et al. 2014) trained on Twitter and Wikipedia, word2vec embeddings (Mikolov et al. 2013) trained on Google News, and FastText (Mikolov et al. 2018) embeddings trained on multiple sources. In addition, we also experiment with removing stop-words from the average word representation, as stop-words can dominate in the average and result in less meaningful sentence representation. For Arabic, we use word2vec embeddings that are trained on Arabic tweets and Arabic Wikipedia (Soliman et al. 2017). Another way to extract contextual features is to use BERT (Devlin et al. 2019) embeddings that are trained using the context of the word in a sentence. BERT is usually trained on a very large text corpus which makes it very useful for off-the-shelf feature extraction and fine-tuning for downstream tasks in NLP. To get one embedding per tweet, we follow the observations made in Devlin et al. (2019) that different layers of BERT capture different kinds of information, so an appropriate pooling strategy should be applied depending on the task. The paper also suggests that the last four hidden layers of the network are good for transfer learning tasks and thus we experiment with 4 different combinations, i.e., concatenate the last 4 hidden layers, average of last 4 hidden layers, last hidden layer and 2nd last hidden layer. We normalise the final embedding so that $l2$ norm of the vector is 1. We also experimented with BERT's pooled sentence embedding that is encoded in the *CLS* (class) tag, which performed significantly more poorly than the pooling strategies we employed. For Arabic, we only experimented with a sentence-transformer (Reimers and Gurevych 2019) that is trained on a multilingual training corpus and outputs a sentence embedding for each tweet/sentence.

To get the overall representation of the tweet, we concatenate all the syntactic features together with either average word embedding or BERT-based transformer features and then apply PCA for dimensionality reduction. An SVM classifier is trained on the feature vectors of tweets to output a binary decision (check-worthy or not). The overall process is shown in Fig. 11.2.

11.3.2 Experiments and Results

In this subsection, we present the important step of pre-processing text, dataset and training setup details, and the evaluation of our proposed model compared to other top models in the challenge.

11.3.2.1 Pre-processing

We use two publicly available pre-processing tools for English and Arabic tweets. We use Baziotis et al.'s tool (Baziotis et al. 2017) for English to apply the following normalisation steps: tokenisation, lower-casing, removal of punctuation, spelling correction, normalise *hashtags, all-caps, censored, elongated* and *repeated* words, and terms like *URL, email, phone, user mentions*. We use the Stanford Stanza toolkit (Qi et al. 2020) to pre-process Arabic tweets by applying the following normalisation steps: tokenisation, multi-word token expansion and lemmatisation. We use the pre-processed text and run off-the-shelf tools to extract syntactic information from tweets and then convert each group of information to feature sets. For English we used spaCy (Honnibal et al. 2020), and Stanford Stanza (Qi et al. 2020) for Arabic tweets, to extract the syntactic features. In all the features, we experiment with keeping and removing stop-words to evaluate their affect. In the case of extracting word embeddings from a transformer network, we use the raw text as the networks have their own tokenisation process.

11.3.2.2 Dataset and Training Details

The English dataset consists of training, development (dev) and test splits, with 672, 150 and 140 tweets respectively on the topic of COVID-19. For validation purposes, we keep 10% (150 samples) from the training data as a development set. We perform grid search using the development set to find the best parameters. The Arabic dataset consists of training and test splits, with 1500 tweets on 3 topics and 6000 tweets on 12 topics respectively, with 500 tweets on each topic. The Arabic dataset consists of tweets on multiple events and topics, which makes the problem harder than in the English dataset. Tweets in the training data are on three topics, "protests in lebanon", "Waseem Youssef" (a preacher from UAE) and "Turkey's intervention in Syria". On the contrary, tweets in the test set are on entirely different topics, some of which are "the deal of the century", "The Houthis in Yemen", "COVID-19", "feminists", "Sudan and normalisation", "Bidoon case in Kuwait" and "boycott countries and spreading rumours against Qatar". The metrics used for evaluation are based on Mean Average Precision (MAP) and Precision@30 (P@30) for English and Arabic datasets, respectively. These metrics were used in the CLEF Checkthat! 2020 challenge for ranking the approaches from different teams.

To train the SVM models for both English and Arabic, we perform grid search over PCA energy (%) conservation, regularisation parameter C and RBF kernel's *gamma*. The parameters range for PCA varies from 100% (original features) to 95% with decrements of 1, and both C and *gamma* vary between -3 to 3 on a log-scale with 30 steps. For faster training on a large grid search, we use ThunderSVM (Wen et al. 2018) which takes advantage of a GPU or a multi-core system to speed up SVM training.

11.3.2.3 Results

We summarise the results based on the best models that we obtained from the grid search, which are discussed below.

English Dataset Our best model (Model-1) and Model-3 use sentence embeddings computed from BERT-large word embeddings as discussed in Sect. 11.3.1.2. In addition, both submissions use POS tag and dependency relation features. Interestingly, we found that the best performing sentence embeddings did not include stop-words. Model-1 uses an ensemble of predictions from three models trained on concatenated last 4 hidden layers, average of last 4 hidden layers and $2nd$ last hidden layer. Model-3 uses predictions from the model trained on the best performing sentence embedding computed from concatenating last 4 hidden layers. Our Model-2 uses an ensemble of predictions from three models trained with GloVe (Pennington et al. 2014) on Twitter with 25-, 50- and 100-dimensional embeddings but with the same POS tag and dependency relation features. We use majority voting to get the final prediction and the mean of decision values to get the final decision value. We found that removing the stop-words to compute the average of word embeddings actually degraded the performance and hence included them in the average.

We also add some additional results to see the effect of stop-words, POS tags, named entities, dependency relations and ensemble predictions in Table 11.2. The effect of stop-words can be clearly seen in alternative runs of Model-1 and Model-3, where the MAP clearly drops by 1-2 points. Similarly, the negative effect of removing POS tag and dependency relation features can be seen in the rest of the

Table 11.2 Check-worthiness estimation English results, MAP (Mean Average Precision), DRel (dependency relations), NE (named entities)

Models	Stopwords	Ensemble	POS	DRel	NE	Embedding	MAP
Model-1		✓	✓	✓		BERT	**0.7217**
Model-2	✓	✓	✓	✓		GloVe	0.6249
Model-3			✓	✓		BERT	0.7139
Model-1-1	✓	✓	✓	✓		BERT	0.7102
Model-1-2		✓	✓			BERT	0.6965
Model-1-3		✓				BERT	0.7094
Model-1-4		✓	✓	✓	✓	BERT	0.7100
Model-3-1	✓		✓	✓		BERT	0.6889
Model-3-2			✓			BERT	0.7074
Model-3-3						BERT	0.6981
Model-3-4			✓	✓	✓	BERT	0.6940
Williams et al. (2020)	–	–	–	–	–	–	0.8064
Nikolov et al. (2020)	–	–	–	–	–	–	0.8034
Alkhalifa et al. (2020)	–	–	–	–	–	–	0.7141

Best results in bold

Table 11.3 Check-worthiness estimation Arabic results, P@K (Precision@K) and AP (Average Precision)

Models	P@5	P@15	P@30	AP
Model-1	**0.6000**	**0.5944**	**0.5778**	**0.4949**
Model-2	0.5500	0.5611	0.5361	0.4649
Model-3	0.4000	0.4167	0.4472	0.4279
Williams et al. (2020)	0.7333	0.7167	0.7000	0.6232
Kartal and Kutlu (2020)	0.7000	0.7000	0.6444	0.5816
Hasanain and Elsayed (2020)	0.6833	0.6667	0.6417	0.5511

Best results in bold

alternative runs. Lastly, adding named entity features to the original submissions also decreases the precision by 1-2 points. This might be because the tweets have very few named entities and they are not useful to distinguish between check-worthy and not check-worthy claims. For comparison with other teams in the challenge, we show the top 3 results at the bottom of the table for reference. Team Accenture (Williams et al. 2020) fine-tuned a RoBERTa model with an extra mean pooling and a dropout layer to prevent overfitting. Team Alex (Nikolov et al. 2020) experimented with different tweet pre-processing techniques and various transformer models, together with logistic regression and SVM. Their main submission used logistic regression trained on 5-fold predictions from RoBERTa concatenated with tweet metadata. Team QMUL-SDS (Alkhalifa et al. 2020) fine-tuned a BERT model pre-trained specifically on COVID twitter data.

Arabic Dataset Our best performing model (Model-1) uses 100-dimensional word2vec Arabic embeddings trained on a Twitter corpus (Soliman et al. 2017) in combination with POS tag features. Model-2 and Model-3 are redundant in terms of feature use, so we only discuss Model-2 here. In addition to features used in Model-1, it uses dependency relation features and 300-dimensional Twitter embeddings instead of 100-dimensional. Our Model-3 uses only a pre-trained multilingual sentence-transformer[1] (Reimers and Gurevych 2020) that is trained on 10 languages including Arabic. We removed the stop-words from all the features, as keeping them resulted in a poorer performance. *Precision@K* and *Average Precision* (AP) results on the test set are shown in the same order in Table 11.3. For comparison with other teams in the challenge, we show the top 3 results at the bottom of the table for reference. Team Accenture (Williams et al. 2020) experimented with and fine-tuned three different pre-trained Arabic BERT models and used external data to increase the positive instances. Team TOBB-ETU (Kartal and Kutlu 2020) used logistic regression and experimented with Arabic BERT and word embeddings together to classify tweets. Team UB_ET (Hasanain and Elsayed 2020) used a multilingual BERT for ranking tweets by check-worthiness.

[1] https://github.com/UKPLab/sentence-transformers.

11.4 Multimodal Claim Detection

Claims about images were first studied by Zlatkova et al. (2019), analysing the relationship between a claim and the corresponding image. Later Cheema et al. (2020b) extended the text-based claim detection datasets from Barrón-Cedeño et al. (2020) and Gupta et al. (2021) with images, and benchmarked several multimodal models to see whether including an image with the claim text shows any improvement in claim detection. Even though the claim datasets were labelled only on the basis of text, one out of three datasets showed slightly better performance in the multimodal setting.

In this section, we introduce and summarise a multimodal claim detection dataset (Cheema et al. 2022) that is specifically curated and annotated based on image-text tweet pairs. The dataset is a collection of image-text tweets in English and on three topics, COVID-19, Climate change and Technology. Next, we provide details about task description, data crawling, annotation questions and description of the final dataset. The finer details on annotation guidelines and annotation process can be referred to in Cheema et al. (2022).

11.4.1 MM-Claims Dataset

Given a tweet with a corresponding image, the task is to identify important factually verifiable or check-worthy claims. In contrast to related work, we introduce a novel dataset for claim detection that is labelled based on both the tweet and the corresponding image, making the task truly multimodal. Our scope of claims is motivated by Alam et al. (2021) and Gupta et al. (2021), which have provided detailed annotation guidelines. We restrict our dataset to factually verifiable claims (as in Alam et al. (2021)) since these are often the claims that need to be prioritised for fact-checking or verification to limit the spread of misinformation. On the other hand, we also include claims that are personal opinions, comments or claims existing at sub-sentence or sub-clause level (as in Gupta et al. (2021)), with the condition that they are factually verifiable. Subsequently, we extend the definition of claims to images along with factually verifiable and check-worthy claims.

11.4.1.1 Data Collection

In previous work on claim detection in tweets, most of the publicly available English language datasets (Alam et al. 2021; Barrón-Cedeño et al. 2020; Gupta et al. 2021; Nakov et al. 2021) are text-based and on a single topic such as *COVID-19*, or *U.S. 2016 Elections*. To make the problem interesting and broader, we have collected tweets on three topics, *COVID-19*, *Climate Change* and broadly *Technology*, that might be of interest to a wider research community.

We have used an existing collection of tweet IDs, where some are topic-specific Twitter dumps, and extracted tweet text and the corresponding image to create a novel multimodal dataset.

COVID-19 We combined tweets from three Twitter resources (Banda et al. 2020; Dimitrov et al. 2020; Lamsal 2023) that were posted between October 2019 and April 2020. In our dataset, we use tweets in the period from March to April 2020.

Climate Change We used a Twitter resource (Littman and Wrubel 2019) that contains tweet IDs related to climate change from September 2017 to May 2019. The tweets were originally crawled based on hashtags like *climatechange, climatechangeisreal, actonclimate, globalwarming, climatedeniers, climatechangeisfalse*, etc.

Technology For the broad topic of *Technology*, we used the *TweetsKB* (Fafalios et al. 2018) corpus. To avoid the extraction of all the tweets from 2019 to 2020 irrespective of the topic, we followed a two-step process to find tweets remotely related to technology. The corpus is available in the form of RDF (Resource Description Framework) triples with attributes like tweet ID, hashtags, entities and emotion labels, but without tweet text or media content details. First, we selected tweet IDs based on hashtags and entities, and only kept those that contain keywords like *technology, cryptocurrency, cybersecurity, machine learning, nano technology, artificial intelligence, IOT, 5G, robotics, blockchain*, etc. The second step of filtering tweets based on a selected set of hashtags for each topic is described in the next subsection.

From the above resources, we collected a total of 660,492 tweets that are filtered to remove inconsistent samples. The specifics of filtering are provided in detail in Cheema et al. (2022). In summary, we end up with 17,771, 4874, and 62,887 tweets with images for *COVID-19*, *Climate Change* and *Technology*, respectively.

11.4.1.2 Annotation

Here, we provide definitions for all investigated claim aspects, the questions asked to annotators, and the cues and explanations for the annotation questions. We define a claim as to *state or assert that something is the case, typically without providing evidence or proof*, using the definition in the Oxford dictionary, like Gupta et al. (2021).

The definition of a *factually verifiable claim* is restricted to claims that can possibly be verified using external sources. These external sources can be reliable websites, books, scientific reports, scientific publications, credible fact-checked news reports, reports from credible organisations like the World Health Organisation or United Nations. Although we did not provide external links of reliable sources for the content in the tweet, we highlighted named entities that pop-up with the text and image description. External sources are not important at this stage because we are only interested in marking claims, which possibly have incorrect details and information. A list of identifiable cues extended from Barrón-Cedeño et al. (2020) for factually verifiable claims is provided in detail in Cheema et al. (2022).

To define check-worthiness, we follow Barrón-Cedeño et al. (2020) and identify claims as check-worthy if the information in the tweet is, (1) *harmful* (attacks a person, organisation, country, group, race, community, etc.), or (2) *urgent or breaking news* (news-like statements about prominent people, organisations, countries and events), or (3) *up-to-date* (referring to recent official documents with facts, definitions and figures). A detailed description of these cases is provided in Cheema et al. (2022). Given these key points, the answer to whether the claim is check-worthy is subjective since it depends on the person's (annotator's) background and knowledge.

Annotation Questions Based on the definitions above, we decided on the following annotation questions in order to identify factually verifiable claims in multimodal data.

- Q1: *Does the image-text pair contain a factually verifiable claim?—Yes/No*
- Q2: *If "Yes" to Q1, Does the claim contain harmful, up-to-date, urgent or breaking-news information?—Yes/No*
- Q3: *If "Yes" to Q1, Does the image contain information about the claim or the claim itself (in the overlaid text)?—Yes/No*

Question 3 (Q3) intends to identify whether the visual content contributes to a tweet having factually verifiable claims. The question is answered "Yes" if one of the following cases hold true: (1) there exists a piece of evidence (e.g. an event, action, situation or a person's identity) or illustration of certain aspects in the claim text, or (2) the image contains overlay text that itself contains a claim in a text form.

Please note that we asked the annotators to label tweets with respect to the time they were posted. During our annotation dry runs, we observed that there were several false annotations for the tweets where the claims were false but already well known facts. This aspect intends to ignore the veracity of claims since some of the claims become facts over time. In addition, we ignore tweets that are questions and label them as not claims unless the corresponding image consists of a response to the question and is a factually verifiable claim. Exactly three annotators labelled each sample, and we used a majority vote to obtain the final label.

11.4.1.3 Final Dataset

We selected a total of 3400 tweets for manual annotation of training (annotated by external annotators) and evaluation (annotated by internal experts) splits. Each split contains an equal number of samples for the topics: *COVID-19*, *Climate Change* and *Technology*.

Labels for three types of claim[2] annotations are derived:

- binary claim classes: *not a claim*, and *claim*
- tertiary claim classes: *not a claim, claim but not check-worthy*, and *check-worthy claim*
- visual claim classes: *not a claim, visually irrelevant claim*, and *visually relevant claim*

While a majority is always possible for the binary claim classification that allows us to derive unambiguous labels, entirely different labels could be chosen for the tertiary and visually relevant claim classification task since the annotators assign three possible classes. Consequently, it is not possible to derive a label with majority voting when each annotator selects a different option. In Cheema et al. (2022), two sets of splits and experiments are conducted for unambiguous annotated samples and conflict resolved samples, respectively. In this chapter, we focus on unambiguous annotated samples split for simplicity and additional analysis on image-text relationship semantics.

As a result, the *Multimodal Claims (MM-Claims)* dataset[3] discussed in this chapter consists of 2555 (T (training)) and 525 (E (evaluation)) samples. We divided the training set (T) in each case further into training and validation in a 90:10 split for hyper-parameter tuning.

11.4.2 Experiments and Results

In this section, we describe the features and baseline models, and summarise the main experimental results using our novel dataset. We test a variety of features and recent multimodal state-of-the-art models.

11.4.2.1 Feature Extraction

Here, we describe all the image, text and multimodal features used in the experiments. For images, we use the standard pre-processing of resizing and normalising an image, whereas text is cleaned and normalised with the same approach as in Sect. 11.3. Besides digits and alphabets, we also keep punctuation to reflect the syntax and style of a written claim.

---

[2] Here claim is a factually-verifiable claim not any claim.

[3] Source code is available at: https://github.com/TIBHannover/MM_Claims Dataset (Tweet IDs) and labels are available at: https://data.uni-hannover.de/dataset/mm_claims. For complete labelled data access (Images and Tweets), please contact at *gullal.cheema@tib.eu* or *gullalcheema@gmail.com*.

Image Features For image encoding, we use a *ResNet-152* (He et al. 2016) model trained on *ImageNet* (Russakovsky et al. 2015) and extract the 2048-dimensional feature vector from the last pooling layer.

Text Features For encoding tweet and OCR text, we test *BERT* (Devlin et al. 2019) uncased models to extract contextual word embeddings. For classification using Support Vector Machine (SVM, (Cortes and Vapnik 1995)), we employ a pooling strategy by adding the last four layers' outputs and then average them to obtain the final 768-dimensional vector.

Multimodal Features We use the following two pre-trained image-text representation learning architectures to extract multimodal features.

The ***ALBEF*** (ALign BEfore Fuse) embedding (Li et al. 2021) results from a recent multimodal state-of-the-art model for vision-language downstream tasks. It is trained on a combination of several image captioning datasets ($\sim$14 million image-text pairs) and uses *BERT* and a visual transformer (Dosovitskiy et al. 2021) for text and image encoding, respectively. It produces a multimodal embedding of 768 dimensions.

The ***CLIP*** (Contrastive Language-Image Pretraining) model (Radford et al. 2021) is trained without any supervision on 400 million image-text pairs. We evaluate several image encoder backbones including *ResNet* and vision transformer (Dosovitskiy et al. 2021). The *CLIP* model outputs two embeddings of the same size, i.e., the image ($CLIP_I$) and the text ($CLIP_T$) embedding, while $CLIP_{I \oplus T}$ denotes the concatenation of two embeddings.

11.4.2.2 Baselines

In the following, we describe training details, hyper-parameters, input combinations and different baseline models' details.

SVM To obtain unimodal and multimodal embeddings for our experiments, we first use PCA (Principal Component Analysis) to reduce the feature size and train an SVM model with the *RBF* kernel. We perform grid search over PCA energy (%) conservation, regularisation parameter *C* and *RBF* kernel's *gamma*. The parameter range for PCA varies from 100% (original features) to 95% with decrements of 1. The parameter range for *C* and *gamma* varies between -1 to 1 on a log-scale with 15 steps. For multimodal experiments, image and text embeddings are concatenated before passing them to PCA and SVM. We normalise the final embedding so that *l2* norm of the vector is 1.

BERT and ALBEF Fine-Tuning (FT) We experiment with fine-tuning the last few layers of unimodal and multimodal transformer models to get a strong multimodal baseline and see whether introducing cross-modal interactions improves claim detection performance. We fine-tune the last layers of both the models and report the best ones in Table 11.4. Detailed additional experimental results on fine-tuned layers can be referred to in Cheema et al. (2022).

Table 11.4 Accuracy (Acc) and Macro-F1 (F1) for binary (BCD) and tertiary claim detection (TCD) in percent [%]. Unless FT (fine-tuning) is written, all models (except MVAE and SpotFake) are SVM models trained on extracted features

Task →	BCD		TCD	
Models ↓	Acc	F1	Acc	F1
Random	50.7	50.2	33.7	28.2
Majority	62.7	38.5	62.7	38.5
ImageNet	63.1	62.6	62.5	40.9
CLIP_I	70.0	69.8	68.9	50.2
BERT	80.5	79.9	77.9	52.9
↪ FT	80.9	80.1	78.3	51.2
CLIP_T	75.6	74.7	77.3	54.4
BERT ⊕ ImageNet	**81.4**	80.9	77.5	56.0
↪ ⊕ OCR	80.9	80.4	77.7	55.0
$\text{CLIP}_{I \oplus T}$	77.8	77.4	77.5	56.4
CLIP_I ⊕ BERT	80.3	79.7	77.9	53.3
ALBEF	76.9	76.5	76.6	55.0
↪ FT	80.2	79.7	**80.0**	63.3
↪ ⊕ OCR ⊕ FT	**81.4**	**81.1**	78.7	**63.5**
MVAE	64.1	62.9	62.9	43.2
SpotFake	71.8	71.4	70.7	50.4

Best results in bold

Models with OCR Text To incorporate OCR text embeddings into our models, we experiment with two strategies for embedding generation and one strategy to fine-tune models. To obtain an embedding for SVM models, we experimented with concatenating the OCR embedding to image and tweet text embeddings as well as adding the OCR embedding directly to tweet text embedding. To fine-tune the models, we concatenate the OCR text to tweet text and limit the OCR text to 128 tokens.

Hyper-parameter Details For fine-tuning, we limit the tweet text to the maximum number of tokens (91) seen in a tweet in the training data and pad the shorter tweets with zeros. For fine-tuning *BERT* and *ALBEF*, we use a batch size of 16 and 8 (size constraints), respectively. We train the models for five epochs and use the best performing model (in terms of accuracy on the validation set) for evaluation. For *BERT*, a dropout with the ratio of 0.2 is applied before the classification head. Further, we use AdamW (Loshchilov and Hutter 2019) as the optimiser with a learning rate of 3*e*-5 and a linear warm-up schedule. The learning rate is first linearly increased from 0 to 3*e*-5 for iterations in the first epoch and then linearly decreased to 0 for the rest of the iterations in 4 epochs. For *ALBEF*, we use the recommended fine-tuning hyper-parameters and settings from the publicly available code.

11.4.2.3 State-of-the-Art Baselines

We compare our models with two state-of-the-art approaches for multimodal fake news detection.

MVAE (Khattar et al. 2019) is a multimodal variational auto-encoder model that uses a multi-task loss to minimise the reconstruction error of individual modalities and task-specific cross-entropy loss for classification. We use the publicly available source code and hyper-parameters for our task.

SpotFake (Singhal et al. 2019) is a model built as a shallow multimodal neural network on top of *VGG-19* image and *BERT* text embeddings using a cross-entropy loss. We re-implement the model in PyTorch and use the hyper-parameter settings given in the paper.

11.4.2.4 Results

We report accuracy (Acc) and Macro-F1 (F1) for binary (BCD) and tertiary claim detection (TCD) in Table 11.4. We also present the fraction (in %) of visually relevant and visually irrelevant (textual only) claims retrieved by each model in Table 11.5. Although we do not train the models specifically to detect visual claim labels, we analyse the fraction of retrieved samples in order to evaluate the bias of binary classification models towards a modality.

Results for Unimodal Models For image-based models, $CLIP_I$ performs (70.0, 69.8) considerably better than *ResNet-152*'s *ImageNet* (63.1, 62.6) features in terms of both accuracy and F1 metrics (Table 11.4, block 2). This result is compliant to previous work (Kirk et al. 2021) where the task has a variety of information and text in images. It is further exaggerated and clearly observable in Table 11.5 where the fraction of visually relevant claims retrieved using $CLIP_I$ (70.3) is higher and comparable to fine-tuned *ALBEF* ⊕ OCR (71.2).

For text-based models, fine-tuning (FT) *BERT* gives the best performance, better than any other unimodal model. This result indicates that the problem is inherently a text-dominant task. SVM trained on BERT features also retrieves the most visually

Table 11.5 Visually-relevant (V) and visually-irrelevant (T) claim detection evaluation. The amount of test samples is reported in brackets and the fraction, how many of them were retrieved, is given in percent [%]. The underlying models are trained for binary claim detection (BCD). The labels for visual relevance are only used for retrieval evaluation

Models	V (76)	T (120)
ImageNet	39.8	39.2
$CLIP_I$	72.4	69.2
BERT	52.6	80.0
↪ FT	53.9	79.2
$CLIP_T$	51.3	76.7
BERT ⊕ ImageNet	63.2	68.3
↪ ⊕ OCR	69.7	78.3
$CLIP_{I \oplus T}$	68.4	70.0
$CLIP_I \oplus BERT$	60.5	75.0
ALBEF	63.2	77.5
↪ FT	65.8	79.2
↪ ⊕ OCR ⊕ FT	**76.3**	**82.5**

Best results in bold

irrelevant claims. It should be noted that textual models can still identify visually relevant claims since they can have a claim or certain cues in the tweet text that refer to the image. Finally, the $CLIP_T$ features perform considerably worse than *BERT* features on BCD, possibly because *CLIP* is limited to short text (75 tokens) and is not trained like vanilla *BERT* on a large text corpus.

Results for Multimodal Models For multimodal models the combination of *BERT* and *ResNet-152* features performs slightly better (0.5–1%) on two metrics in Table 11.4 on the full dataset in a binary task. Although this gain is not impressive, the benefit of combining two modalities is more obvious in identifying visually relevant claims ($>$ 10%) in Table 11.5, which comes at the cost of a lower fraction of visually irrelevant claims. Similarly with *CLIP*, the combination of image and text features ($\text{CLIP}_{I \oplus T}$) improves the overall accuracy from CLIP_I or CLIP_T. However, we do not see the same result for identifying visually relevant claims ($<$ 4–5%). We also experiment with the combination of *BERT* features with *CLIP*'s image features, which improves the overall accuracy further but indicates that the model relies strongly on text (68.4 vs. 60.5 visual retrieval %) rather than the combination. The stronger reliance on text is possibly not a trait of the model alone, but could also be caused by an incompatibility of *BERT* and CLIP_I features.

Finally, we achieve the best performance (by 1–4%) on binary and tertiary claim detection by fine-tuning the *ALBEF* with and without OCR, respectively (Table 11.4, block 3, last two rows). While the benefit of using OCR text in SVM models is not optimal and not considerably helpful, OCR addition to *ALBEF* retrieves the maximum number of visually relevant (76.3%) and visually irrelevant claims (82.5%) (Table 11.5, block 2, last row). These results point towards a major challenge of combining multiple modalities and retaining intra-modal information (and influence) for the task at hand. Figure 11.3 shows a few examples where our best multimodal model correctly classifies, whereas unimodal models based on either image or text do not. All the samples in the figure have images that have some connection to the tweet text. The image in Fig. 11.3b has a connection to one of the words or phrases (e.g., washing your hands) in the tweet text but is not relevant for the claim itself. Figure 11.3a includes an image with the claim itself and a very generic scene in the background. Both image and text in Fig. 11.3c and d are relevant, and the image acts as evidence and additional information. In all these examples, a rich set of information extraction and complex cross-modal learning is required to identify claims in multimodal tweets. When comparing results of recent state-of-the-art architectures for fake news detection, SpotFake (Singhal et al. 2019) does considerably better than MVAE (Khattar et al. 2019) but worse than any of our baseline models.

Fig. 11.3 Qualitative examples where our best multimodal model classifies correctly and unimodal models do not. *F*—false classification, *T*—true classification. **Due to possible licensing issues, best alternative images are shown here*

11.5 Conclusions and Future Work

In this chapter, we presented a text-based approach and a multimodal dataset for claim detection on social media. First, for the text-based approach (Cheema et al. 2020a), we used syntactic and contextual features for predicting the check-worthiness of tweets in Arabic and English with an SVM classifier. For syntactic features, we evaluated parts-of-speech tags, named entities and syntactic dependency relations and used the best feature sets for both languages. In the case of contextual features, we evaluated different word embeddings, and BERT models to capture the semantics of each tweet. Second, we presented a novel *MM-Claims* dataset to foster research on multimodal claim analysis first introduced in Cheema et al. (2022). The dataset has been curated from Twitter data and contains more than 3000 manually annotated tweets for three tasks related to claim detection across three topics, *COVID-19*, *Climate Change* and *Technology*. We evaluated several baseline approaches and compared them against two state-of-the-art fake news detection approaches. Our experimental results suggest that the fine-tuning of pre-trained multimodal and unimodal architectures such as *ALBEF* and *BERT* yield the best performance. We also observed that the overlaid text in images is important in information dissemination, particularly for claim detection. To this end, we evaluated a couple of strategies to incorporate OCR text into our models, which yielded a much better trade-off between identifying visually relevant and visually irrelevant (text-only) claims.

An extended version of the multimodal claims dataset with recent tweets from 2021 and 2022 was introduced in the CLEF Checkthat! 2023 challenge (Barrón-Cedeño et al. 2023). In the future, we will explore other novel architectures for multimodal representation learning and other information extraction techniques to incorporate individual modalities more effectively. With recent development in autoregressive GPT-like multimodal chain-of-thought and reasoning models, it will be interesting to analyse whether these models can also understand complex (e.g., infographics, graphs, text-in-image) multimodal claims.

Acknowledgments This work was funded by European Union's Horizon 2020 research and innovation programme under the Marie Skłodowska-Curie grant agreement no 812997 (CLEOPATRA project), and by the German Federal Ministry of Education and Research (BMBF, FakeNarratives project, no. 16KIS1517)

References

Ajao O, Bhowmik D, Zargari S (2018) Fake news identification on twitter with hybrid CNN and RNN models. In: International Conference on Social Media and Society, SMSociety 2018, Copenhagen, Denmark, July 18–20, 2018, ACM, pp 226–230. https://doi.org/10.1145/3217804.3217917

Alam F, Shaar S, Dalvi F, Sajjad H, Nikolov A, Mubarak H, Martino GDS, Abdelali A, Durrani N, Darwish K, Al-Homaid A, Zaghouani W, Caselli T, Danoe G, Stolk F, Bruntink B, Nakov P (2021) Fighting the COVID-19 infodemic: modeling the perspective of journalists, fact-checkers, social media platforms, policy makers, and the society. In: Conference on Empirical Methods in Natural Language Processing, EMNLP 2021, 16–20 November 2021, Virtual Event, Association for Computational Linguistics, pp 611–649. https://doi.org/10.18653/V1/2021.FINDINGS-EMNLP.56

Albahar MA (2021) A hybrid model for fake news detection: leveraging news content and user comments in fake news. IET Inf Secur 15(2):169–177. https://doi.org/10.1049/ise2.12021

Alkhalifa R, Yoong T, Kochkina E, Zubiaga A, Liakata M (2020) QMUL-SDS at checkthat! 2020 determining COVID-19 tweet check-worthiness using an enhanced CT-BERT with numeric expressions. In: Working Notes of CLEF 2020 - Conference and Labs of the Evaluation Forum, September 22–25, 2020, Virtual Event, CEUR-WS.org, CEUR Workshop Proceedings, vol 2696. https://ceur-ws.org/Vol-2696/paper_186.pdf

Banda JM, Tekumalla R, Wang G, Yu J, Liu T, Ding Y, Chowell G (2020) A large-scale COVID-19 twitter chatter dataset for open scientific research - an international collaboration. CoRR abs/2004.03688. https://arxiv.org/abs/2004.03688. 2004.03688

Barrón-Cedeño A, Elsayed T, Nakov P, Martino GDS, Hasanain M, Suwaileh R, Haouari F, Babulkov N, Hamdan B, Nikolov A, Shaar S, Ali ZS (2020) Overview of checkthat! 2020: automatic identification and verification of claims in social media. In: Experimental IR Meets Multilinguality, Multimodality, and Interaction - 11th International Conference of the CLEF

Association, CLEF 2020, September 22–25, 2020, Virtual Event, Springer, Lecture Notes in Computer Science, vol 12260, pp 215–236. https://doi.org/10.1007/978-3-030-58219-7_17

Barrón-Cedeño A, Alam F, Caselli T, Martino GDS, Elsayed T, Galassi A, Haouari F, Ruggeri F, Struß JM, Nandi RN, Cheema GS, Azizov D, Nakov P (2023) The CLEF-2023 checkthat! lab: checkworthiness, subjectivity, political bias, factuality, and authority. In: European Conference on Information Retrieval, ECIR 2023, Dublin, April 2–6, 2023, Springer, Lecture Notes in Computer Science, vol 13982, pp 506–517. https://doi.org/10.1007/978-3-031-28241-6_59

Baziotis C, Pelekis N, Doulkeridis C (2017) Datastories at semeval-2017 task 4: deep LSTM with attention for message-level and topic-based sentiment analysis. In: International Workshop on Semantic Evaluation co-located with Annual Conference of the North American Chapter of the Association for Computational Linguistics: Human Language Technologies, SemEval@NAACL-HLT 2016, San Diego, CA, June 16–17, 2016, The Association for Computer Linguistics, pp 747–754. https://doi.org/10.18653/V1/S17-2126

Boididou C, Papadopoulos S, Dang-Nguyen D, Boato G, Riegler M, Middleton SE, Petlund A, Kompatsiaris Y (2016) Verifying multimedia use at mediaeval 2016. In: Working Notes Proceedings of the MediaEval 2016 Workshop, Hilversum, The Netherlands, October 20–21, 2016, CEUR-WS.org, CEUR Workshop Proceedings, vol 1739. https://ceur-ws.org/Vol-1739/MediaEval_2016_paper_3.pdf

Cao J, Qi P, Sheng Q, Yang T, Guo J, Li J (2020) Exploring the role of visual content in fake news detection. In: Disinformation, Misinformation, and Fake News in Social Media: Emerging Research Challenges and Opportunities, Springer, Cham, pp 141–161. https://doi.org/10.1007/978-3-030-42699-6_8

Chakrabarty T, Hidey C, McKeown K (2019) IMHO fine-tuning improves claim detection. In: Conference of the North American Chapter of the Association for Computational Linguistics: Human Language Technologies, NAACL-HLT 2019, Minneapolis, MN, June 2–7, 2019, Association for Computational Linguistics, pp 558–563. https://doi.org/10.18653/V1/N19-1054

Cheema GS, Hakimov S, Ewerth R (2020a) Check_square at checkthat! 2020 claim detection in social media via fusion of transformer and syntactic features. In: Working Notes of CLEF 2020 - Conference and Labs of the Evaluation Forum, September 22–25, 2020, Virtual Event, CEUR-WS.org, CEUR Workshop Proceedings, vol 2696. https://ceur-ws.org/Vol-2696/paper_216.pdf

Cheema GS, Hakimov S, Ewerth R (2020b) Tib's visual analytics group at mediaeval '20: detecting fake news on corona virus and 5g conspiracy. In: Working Notes Proceedings of the MediaEval 2020 Workshop, 14–15 December 2020, Virtual Event, CEUR-WS.org, CEUR Workshop Proceedings, vol 2882. https://ceur-ws.org/Vol-2882/paper56.pdf

Cheema GS, Hakimov S, Müller-Budack E, Ewerth R (2021) On the role of images for analyzing claims in social media. In: International Workshop on Cross-lingual Event-centric Open Analytics Co-located with the The Web Conference, CLEOPATRA@WWW 2021, April 12, 2021, Virtual Event, CEUR-WS.org, CEUR Workshop Proceedings, vol 2829, pp 32–46. http://ceur-ws.org/Vol-2829/paper3.pdf

Cheema GS, Hakimov S, Sittar A, Müller-Budack E, Otto C, Ewerth R (2022) Mm-claims: a dataset for multimodal claim detection in social media. In: Conference of the North American Chapter of the Association for Computational Linguistics: Human Language Technologies, NAACL-HLT 2022, Seattle, WA, July 10–15, 2022, Association for Computational Linguistics, pp 962–979. https://doi.org/10.18653/v1/2022.findings-naacl.72

Cortes C, Vapnik V (1995) Support-vector networks. Mach Learn 20(3):273–297. https://doi.org/10.1007/BF00994018

Daxenberger J, Eger S, Habernal I, Stab C, Gurevych I (2017) What is the essence of a claim? Cross-domain claim identification. In: Conference on Empirical Methods in Natural Language Processing, EMNLP 2017, Copenhagen, Denmark, September 9–11, 2017, Association for Computational Linguistics, pp 2055–2066. https://doi.org/10.18653/V1/D17-1218

Devlin J, Chang M, Lee K, Toutanova K (2019) BERT: pre-training of deep bidirectional transformers for language understanding. In: Conference of the North American Chapter of

the Association for Computational Linguistics: Human Language Technologies, NAACL-HLT 2019, Minneapolis, MN, June 2–7, 2019, Association for Computational Linguistics, pp 4171–4186. https://doi.org/10.18653/V1/N19-1423
Dimitrov D, Baran E, Fafalios P, Yu R, Zhu X, Zloch M, Dietze S (2020) Tweetscov19 - A knowledge base of semantically annotated tweets about the COVID-19 pandemic. In: ACM International Conference on Information and Knowledge Management, CIKM 2020, October 19–23, 2020, Virtual Event, ACM, pp 2991–2998. https://doi.org/10.1145/3340531.3412765
Dosovitskiy A, Beyer L, Kolesnikov A, Weissenborn D, Zhai X, Unterthiner T, Dehghani M, Minderer M, Heigold G, Gelly S, Uszkoreit J, Houlsby N (2021) An image is worth 16x16 words: transformers for image recognition at scale. In: International Conference on Learning Representations, ICLR 2021, May 3–7, 2021, Virtual Event, OpenReview.net. https://openreview.net/forum?id=YicbFdNTTy
D'Ulizia A, Caschera MC, Ferri F, Grifoni P (2021) Fake news detection: a survey of evaluation datasets. PeerJ Comput Sci 7:e518. https://doi.org/10.7717/peerj-cs.518
Fafalios P, Iosifidis V, Ntoutsi E, Dietze S (2018) Tweetskb: A public and large-scale RDF corpus of annotated tweets. In: European Semantic Web Conference, ESWC 2018, Heraklion, Crete, Greece, June 3–7, 2018, Springer, Lecture Notes in Computer Science, vol 10843, pp 177–190. https://doi.org/10.1007/978-3-319-93417-4_12
Giachanou A, Rosso P, Crestani F (2019) Leveraging emotional signals for credibility detection. In: International ACM Conference on Research and Development in Information Retrieval, SIGIR 2019, Paris, July 21–25, 2019, ACM, pp 877–880. https://doi.org/10.1145/3331184.3331285
Giachanou A, Zhang G, Rosso P (2020a) Multimodal fake news detection with textual, visual and semantic information. In: International Conference on Text, Speech and Dialogue, TSD 2020, September 8–11, 2020, Virtual Event, Springer, Lecture Notes in Computer Science, vol 12284, pp 30–38. https://doi.org/10.1007/978-3-030-58323-1_3
Giachanou A, Zhang G, Rosso P (2020b) Multimodal multi-image fake news detection. In: IEEE International Conference on Data Science and Advanced Analytics, DSAA 2020, October 6–9, 2020, Virtual Event, IEEE, pp 647–654. https://doi.org/10.1109/DSAA49011.2020.00091
Gupta S, Singh P, Sundriyal M, Akhtar MS, Chakraborty T (2021) LESA: linguistic encapsulation and semantic amalgamation based generalised claim detection from online content. In: Conference of the European Chapter of the Association for Computational Linguistics, EACL 2021, April 19–23, 2021, Virtual Event, Association for Computational Linguistics, pp 3178–3188. https://doi.org/10.18653/V1/2021.EACL-MAIN.277
Hasanain M, Elsayed T (2020) bigir at checkthat! 2020: multilingual BERT for ranking arabic tweets by check-worthiness. In: Working Notes of CLEF 2020 - Conference and Labs of the Evaluation Forum, Thessaloniki, Greece, September 22–25, 2020, CEUR-WS.org, CEUR Workshop Proceedings, vol 2696. https://ceur-ws.org/Vol-2696/paper_142.pdf
He K, Zhang X, Ren S, Sun J (2016) Deep residual learning for image recognition. In: IEEE Conference on Computer Vision and Pattern Recognition, CVPR 2016, Las Vegas, NV, June 27–30, 2016, IEEE Computer Society, pp 770–778. https://doi.org/10.1109/CVPR.2016.90
Honnibal M, Montani I, Van Landeghem S, Boyd A (2020) spaCy: industrial-strength natural language processing in Python. https://zenodo.org/doi/10.5281/zenodo.1212303
Iskender N, Schaefer R, Polzehl T, Möller S (2021) Argument mining in tweets: comparing crowd and expert annotations for automated claim and evidence detection. In: International Conference on Applications of Natural Language to Information Systems, NLDB 2021, Saarbrücken, June 23–25, 2021, Springer, Lecture Notes in Computer Science, vol 12801, pp 275–288. https://doi.org/10.1007/978-3-030-80599-9_25
Jin Z, Cao J, Guo H, Zhang Y, Luo J (2017) Multimodal fusion with recurrent neural networks for rumor detection on microblogs. In: ACM on Multimedia Conference, MM 2017, Mountain View, CA, October 23–27, 2017, ACM, pp 795–816. https://doi.org/10.1145/3123266.3123454
Jindal S, Sood R, Singh R, Vatsa M, Chakraborty T (2020) Newsbag: a benchmark multimodal dataset for fake news detection. In: Workshop on Artificial Intelligence Safety Co-located with AAAI Conference on Artificial Intelligence, SafeAI@AAAI 2020, New York City, NY,

February 7, 2020, CEUR-WS.org, CEUR Workshop Proceedings, vol 2560, pp 138–145. http://ceur-ws.org/Vol-2560/paper27.pdf

Kartal YS, Kutlu M (2020) TOBB ETU at checkthat! 2020: prioritizing english and arabic claims based on check-worthiness. In: Working Notes of CLEF 2020 - Conference and Labs of the Evaluation Forum, Thessaloniki, Greece, September 22–25, 2020, CEUR-WS.org, CEUR Workshop Proceedings, vol 2696. https://ceur-ws.org/Vol-2696/paper_234.pdf

Khattar D, Goud JS, Gupta M, Varma V (2019) MVAE: multimodal variational autoencoder for fake news detection. In: International Conference on World Wide Web, WWW 2019, San Francisco, CA, May 13–17, 2019, ACM, pp 2915–2921. https://doi.org/10.1145/3308558.3313552

Kirk H, Jun Y, Rauba P, Wachtel G, Li R, Bai X, Broestl N, Doff-Sotta M, Shtedritski A, Asano YM (2021) Memes in the wild: assessing the generalizability of the hateful memes challenge dataset. In: Workshop on Online Abuse and Harms Co-located with Annual Meeting of the Association for Computational Linguistics and International Joint Conference on Natural Language Processing, WOAH@ACL-IJCNLP 2021, August 5–6, 2021, Virtual Event, Association for Computational Linguistics, Online, pp 26–35. https://doi.org/10.18653/v1/2021.woah-1.4

Lamsal R (2023) Coronavirus (COVID-19) tweets dataset. https://doi.org/10.21227/781W-EF42

Levy R, Bilu Y, Hershcovich D, Aharoni E, Slonim N (2014) Context dependent claim detection. In: International Conference on Computational Linguistics, COLING 2014, Dublin, August 23–29, 2014, ACL, pp 1489–1500. https://aclanthology.org/C14-1141/

Li J, Selvaraju RR, Gotmare A, Joty SR, Xiong C, Hoi SC (2021) Align before fuse: vision and language representation learning with momentum distillation. In: Annual Conference on Neural Information Processing Systems, NeurIPS 2021, December 6–14, 2021, Virtual Event, pp 9694–9705. https://proceedings.neurips.cc/paper/2021/hash/505259756244493872b7709a8a01b536-Abstract.html

Lippi M, Torroni P (2015) Context-independent claim detection for argument mining. In: International Joint Conference on Artificial Intelligence, IJCAI 2015, Buenos Aires, July 25–31, 2015, AAAI Press, pp 185–191. http://ijcai.org/Abstract/15/033

Littman J, Wrubel L (2019) Climate Change Tweets Ids. https://doi.org/10.7910/DVN/5QCCUU

Loshchilov I, Hutter F (2019) Decoupled weight decay regularization. In: International Conference on Learning Representations, ICLR 2019, New Orleans, LA, May 6–9, 2019, OpenReview.net. https://openreview.net/forum?id=Bkg6RiCqY7

Mikolov T, Sutskever I, Chen K, Corrado GS, Dean J (2013) Distributed representations of words and phrases and their compositionality. In: Annual Conference on Neural Information Processing Systems, NIPS 2013, Lake Tahoe, Nevada, December 5–8, 2013, pp 3111–3119. https://proceedings.neurips.cc/paper/2013/hash/9aa42b31882ec039965f3c4923ce901b-Abstract.html

Mikolov T, Grave E, Bojanowski P, Puhrsch C, Joulin A (2018) Advances in pre-training distributed word representations. In: International Conference on Language Resources and Evaluation, LREC 2018, Miyazaki, May 7–12, 2018, European Language Resources Association (ELRA). http://www.lrec-conf.org/proceedings/lrec2018/summaries/721.html

Nakamura K, Levy S, Wang WY (2020) Fakeddit: a new multimodal benchmark dataset for fine-grained fake news detection. In: International Conference on Language Resources and Evaluation, LREC 2020, May 11–16, 2020, Virtual Event, European Language Resources Association, pp 6149–6157. https://aclanthology.org/2020.lrec-1.755/

Nakov P, Martino GDS, Elsayed T, Barrón-Cedeño A, Míguez R, Shaar S, Alam F, Haouari F, Hasanain M, Babulkov N, Nikolov A, Shahi GK, Struß JM, Mandl T (2021) The CLEF-2021 checkthat! lab on detecting check-worthy claims, previously fact-checked claims, and fake news. In: European Conference on Information Retrieval, ECIR 2021, March 28–April 1, 2021, Virtual Event, Springer, Lecture Notes in Computer Science, vol 12657, pp 639–649. https://doi.org/10.1007/978-3-030-72240-1_75

Nakov P, Barrón-Cedeño A, Martino GDS, Alam F, Struß JM, Mandl T, Míguez R, Caselli T, Kutlu M, Zaghouani W, Li C, Shaar S, Shahi GK, Mubarak H, Nikolov A, Babulkov N, Kartal YS, Wiegand M, Siegel M, Köhler J (2022) Overview of the CLEF-2022 checkthat! lab on fighting the COVID-19 infodemic and fake news detection. In: Experimental IR Meets Multilinguality, Multimodality, and Interaction - 13th International Conference of the CLEF Association, CLEF 2022, Bologna, Italy, September 5–8, 2022, Springer, Lecture Notes in Computer Science, vol 13390, pp 495–520. https://doi.org/10.1007/978-3-031-13643-6_29

Nguyen V, Sugiyama K, Nakov P, Kan M (2022) FANG: leveraging social context for fake news detection using graph representation. Commun ACM 65(4):124–132. https://doi.org/10.1145/3517214

Nikolov A, Martino GDS, Koychev I, Nakov P (2020) Team alex at CLEF checkthat! 2020: identifying check-worthy tweets with transformer models. In: Working Notes of CLEF 2020 - Conference and Labs of the Evaluation Forum, September 22–25, 2020, Virtual Event, CEUR-WS.org, CEUR Workshop Proceedings, vol 2696. https://ceur-ws.org/Vol-2696/paper_170.pdf

Pennington J, Socher R, Manning CD (2014) Glove: global vectors for word representation. In: Conference on Empirical Methods in Natural Language Processing, EMNLP 2014, Doha, October 25–29, 2014, ACL, pp 1532–1543. https://doi.org/10.3115/V1/D14-1162

Qi P, Zhang Y, Zhang Y, Bolton J, Manning CD (2020) Stanza: a python natural language processing toolkit for many human languages. In: Annual Meeting of the Association for Computational Linguistics: System Demonstrations, ACL 2020, July 5–10, 2020, Virtual Event, Association for Computational Linguistics, pp 101–108. https://doi.org/10.18653/V1/2020.ACL-DEMOS.14

Radford A, Kim JW, Hallacy C, Ramesh A, Goh G, Agarwal S, Sastry G, Askell A, Mishkin P, Clark J, Krueger G, Sutskever I (2021) Learning transferable visual models from natural language supervision. In: International Conference on Machine Learning, ICML 2021, 18–24 July 2021, Virtual Event, PMLR, Proceedings of Machine Learning Research, vol 139, pp 8748–8763. http://proceedings.mlr.press/v139/radford21a.html

Reimers N, Gurevych I (2019) Sentence-bert: sentence embeddings using siamese bert-networks. In: Conference on Empirical Methods in Natural Language Processing and International Joint Conference on Natural Language Processing, EMNLP-IJCNLP 2019, Hong Kong, November 3–7, 2019, Association for Computational Linguistics, pp 3980–3990. https://doi.org/10.18653/V1/D19-1410

Reimers N, Gurevych I (2020) Making monolingual sentence embeddings multilingual using knowledge distillation. In: Conference on Empirical Methods in Natural Language Processing, EMNLP 2020, November 16–20, 2020, Virtual Event, Association for Computational Linguistics, pp 4512–4525. https://doi.org/10.18653/V1/2020.EMNLP-MAIN.365

Rosenthal S, McKeown KR (2012) Detecting opinionated claims in online discussions. In: IEEE International Conference on Semantic Computing, ICSC 2012, Palermo, September 19–21, 2012, IEEE Computer Society, pp 30–37. https://doi.org/10.1109/ICSC.2012.59

Russakovsky O, Deng J, Su H, Krause J, Satheesh S, Ma S, Huang Z, Karpathy A, Khosla A, Bernstein MS, Berg AC, Fei-Fei L (2015) Imagenet large scale visual recognition challenge. Int J Comput Vision 115(3):211–252. https://doi.org/10.1007/S11263-015-0816-Y

Shu K, Wang S, Liu H (2019) Beyond news contents: the role of social context for fake news detection. In: ACM International Conference on Web Search and Data Mining, WSDM 2019, Melbourne, VIC, February 11–15, 2019, ACM, pp 312–320. https://doi.org/10.1145/3289600.3290994

Singhal S, Shah RR, Chakraborty T, Kumaraguru P, Satoh S (2019) Spotfake: a multi-modal framework for fake news detection. In: IEEE International Conference on Multimedia Big Data, BigMM 2019, Singapore, September 11–13, 2019, IEEE, pp 39–47. https://doi.org/10.1109/BIGMM.2019.00-44

Soliman AB, Eissa K, El-Beltagy SR (2017) Aravec: a set of arabic word embedding models for use in arabic NLP. In: International Conference On Arabic Computational Linguistics, ACLING 2017, Dubai, November 5–6, 2017, Elsevier, Procedia Computer Science, vol 117, pp 256–265. https://doi.org/10.1016/J.PROCS.2017.10.117

Suykens JAK, Vandewalle J (1999) Least squares support vector machine classifiers. Neural Process Lett 9(3):293–300. https://doi.org/10.1023/A:1018628609742

Wang Y, Ma F, Jin Z, Yuan Y, Xun G, Jha K, Su L, Gao J (2018) EANN: event adversarial neural networks for multi-modal fake news detection. In: ACM International Conference on Knowledge Discovery & Data Mining, SIGKDD 2018, London, August 19–23, 2018, ACM, pp 849–857. https://doi.org/10.1145/3219819.3219903

Wang Y, Qian S, Hu J, Fang Q, Xu C (2020) Fake news detection via knowledge-driven multimodal graph convolutional networks. In: International Conference on Multimedia Retrieval, ICMR 2020, June 8–11, 2020, Virtual Event, ACM, pp 540–547. https://doi.org/10.1145/3372278.3390713

Wen Z, Shi J, Li Q, He B, Chen J (2018) Thundersvm: a fast SVM library on gpus and cpus. J Mach Learn Res 19:21:1–21:5. http://jmlr.org/papers/v19/17-740.html

Williams EM, Rodrigues P, Novak V (2020) Accenture at checkthat! 2020: if you say so: Post-hoc fact-checking of claims using transformer-based models. In: Working Notes of CLEF 2020 - Conference and Labs of the Evaluation Forum, September 22–25, 2020, Virtual Event, CEUR-WS.org, CEUR Workshop Proceedings, vol 2696. https://ceur-ws.org/Vol-2696/paper_226.pdf

Wold S, Esbensen K, Geladi P (1987) Principal component analysis. Chemometr Intell Laborat Syst 2(1):37–52. Proceedings of the Multivariate Statistical Workshop for Geologists and Geochemists. https://doi.org/10.1016/0169-7439(87)80084-9. https://www.sciencedirect.com/science/article/pii/0169743987800849

Wührl A, Klinger R (2021) Claim detection in biomedical twitter posts. In: Workshop on Biomedical Language Processing co-located with Annual Conference of the North American Chapter of the Association for Computational Linguistics: Human Language Technologies, BioNLP@NAACL-HLT 2021, June 11, 2021, Virtual Event, Association for Computational Linguistics, pp 131–142. https://doi.org/10.18653/v1/2021.bionlp-1.15

Zlatkova D, Nakov P, Koychev I (2019) Fact-checking meets fauxtography: Verifying claims about images. In: Conference on Empirical Methods in Natural Language Processing and International Joint Conference on Natural Language Processing, EMNLP-IJCNLP 2019, Hong Kong, November 3–7, 2019, Association for Computational Linguistics, pp 2099–2108. https://doi.org/10.18653/V1/D19-1216

Chapter 12
Narrativising Events

Robert Porzel, Mihai Pomarlan, Laura Spillner, Johanna Rockstroh, and John A. Bateman

Abstract A challenge for automated user access to, or presentation of, event data is the fact that such data rarely explicitly controls for appropriate narrativisation choices. This is problematic because readers nevertheless read narrativisation effects into texts regardless of whether those effects were intended or not. Gaining control of such effects, which we characterise here as the "narrativisation task", therefore constitutes a general problem whenever knowledge is to be externalised for user access, regardless of whether that externalisation is linguistic or involves other presentation modalities. To achieve narrative control in verbalisation, we present a formal theory of narratives using OWL-DL, specifying the theory's ontological commitments and accompanying ontological design patterns. Building on this, we demonstrate a specific pipeline that can be applied to produce linguistic narratives from knowledge graphs via an ontological layer and corresponding rules that generate appropriately slanted natural language from semantic specifications. This complements existing event knowledge by pairing any event representation with a selected conceptualisation, i.e. interpretation. By these means, we aim to improve user access to event information by constructing appropriate narrative stances that help render information presentation coherently in multiple languages; we illustrate this capability here using English and Spanish.

12.1 Introduction

As the map is not the territory, a description of an event is not the event itself. Descriptions of events are narratives that can differ greatly from one another. Spinning a narrative about an event, e.g. the execution of Louis XVI, will entail selecting a subset of entities from the actual events that involved literally innumerable details—from the weeds between the cobblestones and the constellation

R. Porzel (✉) · M. Pomarlan · L. Spillner · J. Rockstroh · J. A. Bateman
Bremen University, Bremen, Germany
e-mail: porzel@uni-bremen.de; pomarlan@uni-bremen.de; laura.spillner@uni-bremen.de; rockstro@uni-bremen.de; bateman@uni-bremen.de

I. Marenzi et al. (eds.), *Event Analytics across Languages and Communities*,
https://doi.org/10.1007/978-3-031-64451-1_12

of the planets at the specific time and place. In the field of narratology, the term *fabula* refers to the actual event as it transpired in the real world, i.e. the facts on the ground, and the *plot* refers to the set of entities and interrelationships selected from the *fabula* to "tell a story". Narratives of events then add to the plot as they introduce perspectives and valence about the event that is being narrativised.

As an example, when describing a war-like situation, it might, theoretically, be possible to say that some governmental head of a country gave an order for the army to move into another country by force. However, natural language renditions of corresponding states of affairs usually contain expressions such as *invading* or *liberating* that assume a specific perspective (taking sides) and denote some valuation of the situation at hand. In other words, rather than objective and neutral event depictions, we are spinning narratives out of the facts on the ground.

While these facts, e.g. as represented by some knowledge graph, may contain large quantities of information, they are, by themselves, not very meaningful. Only when we put them into a pragmatic context do we assign a specific meaning to them. As an illustration, we might interpret a single observed episode as either *throwing something* or *dropping something*. This difference in the narrativisation of an event, consequently, yields two quite distinct narratives:

1. John dropped the glass onto the floor
2. John threw the glass onto the floor

It is important to note that the knowledge graph representing these two minimal narratives could well be identical. We therefore differentiate between a factual knowledge graph, i.e. the *fabula*, which has not been narrativised and a (language-based) description of it, i.e. a *narration*. This pairs a situation with a selected conceptualisation, i.e. interpretation, thereof and provides the basis for rendering the latter in natural language. In addition to becoming meaningful, the description will, in turn, evoke a pragmatic stance that ascribes, e.g. a specific perspective and intention to the agent(s) acting in specific roles within the narrative.

Modelling events and narratives of those events, therefore, poses interesting challenges. In this work, we will outline and motivate the ontological commitments and distinctions pertinent for such a model. This model can, in turn, be employed to generate various narratives about an event that can differ in language as well as perspective and valence. We first introduce the modelling framework and foundational system and then describe how it is employed to construct formal representations of narratives that function as input to a grammar-based generation framework capable of constructing linguistic surface structures corresponding to that input.

12.2 Foundational Framework and Prior Art

The model outlined and discussed herein neither stands alone nor is it the first attempt to cast narratives in a formal framework. Indeed, there is a long history of approaching narrative in computational and formal contexts and for various purposes (Meehan 1980; Lehnert 1982; Elson 2004; Peinado and Pablo Gervás 2004; Swartjes and Theune 2006; Szilas 2007; Zarri 2009; Riedl and Young 2010; Lakoff and Narayanan 2010; Mani 2012; Jasinskaja and Roßdeutscher 2012; Eger et al. 2015). Winer (2014) offers a partial review of earlier work in relation to ontology. Consequently, in this section, we will briefly describe some of the more recent work on formalising narratives before discussing the foundational commitments and ontological framework employed in our own approach.

Each individual approach for representing narratives formally is driven by the specific requirements of the respective application for which it is intended. For example, the model of Meghini et al. (2021) is seeking to organise information provided in digital libraries and thus equates a narrative with an event and allows for events to feature dependent events. The model uses RDF based on OWL, and narrative events can have spatial or temporal relations between them. The main purpose is to connect digitally represented entities to pertinent events, e.g. the Divine Comedy as a book and the person Dante Alighieri can be connected by a narrative *Dante writes the Divine Comedy*. In this approach, *events* are also not formally specified as no foundational framework is employed.

A different approach by Kroll et al. makes a useful distinction between factual relations, as expressed by knowledge graphs, and narrative relations that constitute hypothetical relations connecting factual ones (Kroll et al. 2020). For this, the approach needs to employ RDF* to express relations that range over relations. This approach can, therefore, be employed to postulate, e.g. a causation relation as a narrative that connects hitherto isolated knowledge graphs.

A little closer to the task at hand here is the work described by Evans et al. that seeks to classify (partial) sensory data as a narrative that can be framed as an inductive logic programming task (Evans et al. 2021). Their focus lies on reducing the search space by finding hypotheses—which correspond to narratives in their account—that provide as simple an explanation of the observed data as possible. This approach can, therefore, be deemed compatible, yet orthogonal, to the one presented here, as it does not focus on capturing narrative knowledge explicitly in an ontology but rather offers a classification approach that employs such a model as a target representation.

12.2.1 Foundational Commitments

It has become a sensible and important part of ontology engineering to make the underlying foundational commitments of a given model explicit. Our formal theory of narratives *cum* model is based on the DOLCE+DnS Ultralite (DUL) foundational framework (Gangemi and Mika 2003; Masolo et al. 2003). This decision is strongly motivated by the underlying ontological commitments of DUL, its axiomatisation as well as the incorporation of the *Descriptions and Situations* module. Firstly, DUL is not a revisionary model but seeks to express positions that shape human cognition. Furthermore, DUL allows for a multiplicative approach. However, rather than capturing the flexibility of different understandings of events via multiple inheritance, it is also possible to combine a reduced *ground* classification with a *descriptive* approach for handling event interpretations. For this, a primary branch of the ontology represents the ground **physical model**, e.g. objects and events, while a secondary branch represents the **social model**, e.g. roles and tasks. No entities in the social branch would exist without a cognitive agent, i.e. they constitute social objects that represent concepts about or descriptions of entities, for example, the construal of an observable event where an object moves from an agent's hand to the floor into the interpretations given in Examples 1 and 2.

Every axiomatisation in the physical branch can, therefore, be regarded as expressing some physical context, whereas axiomatisations in the descriptive social branch are used to express social contexts. A set of dedicated relations is provided that connects both branches. For example, the relation *classifies* connects ground objects, e.g. a hammer, with the roles they can play, i.e. potential classifications. Thus, we can state that a hammer can in some context be conceptualised as a murder weapon, a paper weight or a door stopper. Nevertheless, neither its ground ontological classification as a tool will change nor will hammers be subsumed as kinds of door stoppers, paper weights or weapons via multiple inheritance—thus maintaining the essential ontological meta-property of *rigidity* (Guarino and Welty 2004; Welty and Andersen 2005).

12.2.2 The SOMA Ontology

The approach presented here extends the Socio-physical Model of Activities (SOMA) by including a new module for representing narratives. SOMA is based on the DUL foundational framework and its plugin IOLite (Beßler et al. 2020). Consequently, SOMA follows the distinction just introduced between two knowledge branches: one physical and one social. This enforces the distinction between objects and events in the physical branch on the one hand, as well as roles and tasks and descriptions thereof in the social branch on the other. Beßler et al. (2020) propose further that axiomatisations in the physical branch express physical contexts, which can be classified by axiomatisation in the social context. For example, a glass and its properties of being a designed physical artefact would be described using parts of the physical branch, but its potential usage or affordances would be axiomatised using the social branch (Beßler et al. 2020).

SOMA is built out of multiple modules capturing different aspects. For example, the *SAY* module defines the theories required by linguistic processing of instructions

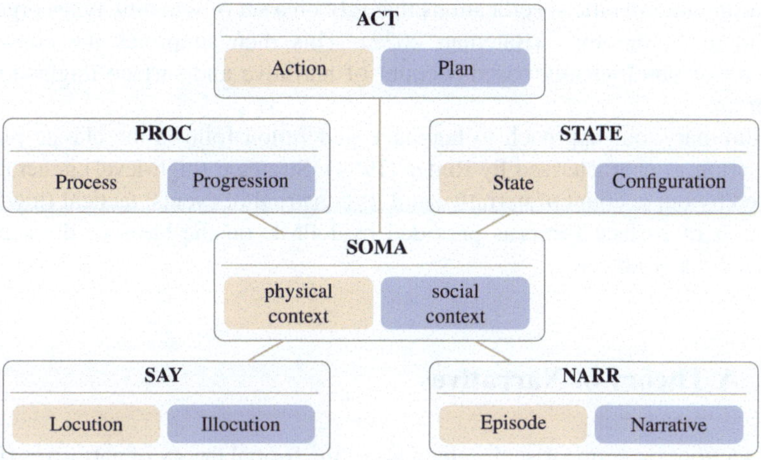

Fig. 12.1 An overview of the SOMA modules

(Porzel and Cangalovic 2020). An overview of some of the modules is given in Fig. 12.1, including the module for representing narratives (NARR) that is the focus of the work described here.

12.2.3 Generating Linguistic Realisations

For generating comprehensible and appropriate natural language expressions, various end-to-end approaches based on some forms of machine learning now exist, both for generating individual sentences based on knowledge graphs (Marcheggiani and Perez-Beltrachini 2018) and for sequences of sentences for sub-graphs retrieved from larger knowledge graphs (Kurisinkel and Chen 2021). The work presented here, however, is closely related to that of constructing formal narrative structures out of distributed knowledge graphs (Blin 2022).

In our approach, therefore, we need to maintain maximal control of the phrasing possibilities offered by a natural language in order to flexibly respond to distinct, but sometimes rather subtle, narrative requirements. To achieve this, we take a symbolic approach leveraging previous work on large-scale multilingual grammar-based natural language generation, employing the Komet-Penman MultiLingual (KPML) system (Bateman 1997) as our natural language generation component. This system offers a well-tested platform for grammar engineering that is specifically designed for natural language generation (Reiter and Dale 2000). KPML employs large-scale grammars written within the framework of Systemic-Functional Linguistics (SFL), which allows us to include linguistic phenomena important for the generation of natural texts and which go beyond the bare propositional content that is to be expressed. In addition, the grammars supplied with KPML are controlled during generation via semantic specifications that rely on a set of semantic types organised as a linguistic ontology (Bateman 2022). This then simplifies the connection between our ontologically based account of narrative and surface linguistic form generation.

In summary, our approach to language generation follows the classic pipeline organisation as characterised by Reiter (1994). Strategic (high-level) generation is provided by our account of narrative and its construal of events; tactical (low-level) realisation of surface forms is provided by KPML on the basis of the semantic specifications produced.

12.3 A Theory of Narratives

In this section, we summarise the elements of the formal theory of narratives (Porzel and Cangalovic 2020) that are pertinent to this endeavour. For this, we first look at the ground partition of the theory, which represents a selective view on events, i.e. the *plot*. In the model, these are considered as *episodes* that occur in the world and can be recorded, e.g. as visual or force dynamic data.[1] These numeric data can be seen as the type of sensory data used in the approach of Evans et al. (2021) described above.

$$\mathtt{Episode}(x) \rightarrow \mathtt{Situation}(x) \tag{12.1}$$

$$\mathtt{Episode}(x) \rightarrow \forall y(INC_E(y,x) \rightarrow \mathtt{Event}(y)) \tag{12.2}$$

$$\mathtt{Episode}(x) \rightarrow \exists y(INC_E(y,x)) \tag{12.3}$$

$$\mathtt{Action}(x) \rightarrow \mathtt{Event}(x) \tag{12.4}$$

$$\mathtt{Event}(x) \rightarrow \forall y(HAS_P(y,x) \rightarrow \mathtt{PhysicalObject}(y)) \tag{12.5}$$

$$\mathtt{Event}(x) \rightarrow \exists y(HAS_P(y,x)) \tag{12.6}$$

We model an *episode* as a dul:situation (12.1) that must include one dul:event; episodes that consist of multiple events are also possible here. Inclusion is denoted by the *includesEvent* (INC_E) relation (12.2, 12.3). As already given by the foundational framework, dul:actions are types of dul:events (12.4), and dul:events have dul:physicalObjects as participants denoted by the dul:hasParticipant (HAS_P)

[1] In the following, all concepts that are part of the SOMA-NARR module will be italicised, and concepts already given by the foundational framework will be denoted by the prefix “dul:”, which is short for dul:<http://ontologydesignpatterns.org/ont/dul/DUL.owl#> where further documentation about these concepts can be found.

relation (12.5, 12.6). An *episode* can now be described by exactly one *narrative* (12.7).

$$\texttt{Episode}(x) \rightarrow \exists y(DESCR(y, x) \wedge \texttt{Narrative}(y)) \tag{12.7}$$

$$DESCR(x, y) \rightarrow \texttt{Description}(x) \tag{12.8}$$

The *describes* relation (*DESCR*) holds between a dul:description (e.g. a *narrative*) and entities that are conceptualised by the description (12.8).

Narratives define construals of dul:tasks and dul:roles within the *episode* they describe. Specifically, we distinguish between dul:tasks that are conceptualisations of dul:actions and the dul:roles that are narrative-specific conceptualisations of the given function of dul:physicalObjects. Consequently, we introduce two relations *definesTask* (DEF_T) and *definesRole* (DEF_R) that link a *narrative* to the respective entities.

$$DEF(x, y) \rightarrow \texttt{Description}(x) \wedge \texttt{Concept}(y) \tag{12.9}$$

$$DEF_T(x, y) \rightarrow DEF(x, y) \wedge \texttt{Narrative}(x) \wedge \texttt{Task}(y) \tag{12.10}$$

$$DEF_R(x, y) \rightarrow DEF(x, y) \wedge \texttt{Narrative}(x) \wedge \texttt{Role}(y) \tag{12.11}$$

Finally, we formalise the *narrative* concept by axiomatising its relationship to the episodic content described and the concepts defined by it.

$$\texttt{Narrative}(x) \rightarrow \texttt{Description}(x) \tag{12.12}$$

$$\texttt{Narrative}(x) \rightarrow \forall y(DESCR(x, y) \rightarrow \texttt{Episode}(y)) \tag{12.13}$$

$$\texttt{Narrative}(x) \rightarrow \exists y(DEF_T(x, y) \wedge \texttt{Task}(y)) \tag{12.14}$$

$$\texttt{Narrative}(x) \rightarrow \exists y(DEF_R(x, y) \wedge \texttt{Role}(y)) \tag{12.15}$$

Our theory views *narratives* as dul:descriptions (12.12) that only describe *episodes* (12.13). A *narrative* defines exactly one dul:task (12.14) and one dul:role (12.15). These relationships between concepts defined in our theory are depicted in Fig. 12.2.

$$NAR(x, y) \rightarrow \texttt{Episode}(x) \wedge \texttt{Concept}(y) \tag{12.16}$$

$$NAR(x, y) \rightarrow \exists a(DESCR(a, x) \wedge DEF(a, y)) \tag{12.17}$$

$$NAR_R(x, y) \rightarrow NAR(x, y) \wedge \texttt{Role}(y) \tag{12.18}$$

$$NAR_R(x, y) \rightarrow \exists a(DESCR(a, x) \wedge DEF_R(a, y)) \tag{12.19}$$

$$NAR_T(x, y) \rightarrow NAR(x, y) \wedge \texttt{Task}(y) \tag{12.20}$$

$$NAR_T(x, y) \rightarrow \exists a(DESCR(a, x) \wedge DEF_T(a, y)) \tag{12.21}$$

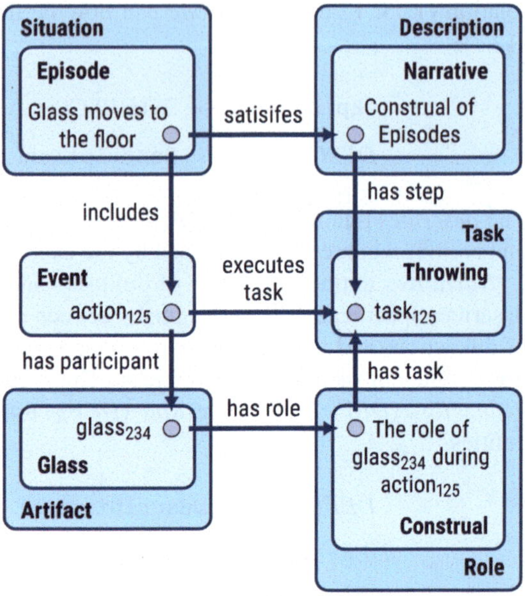

Fig. 12.2 An example of an episode (the physical occurrence) and its narrativisation

The *narrativises* relation links *episodes* and dul:concepts (12.16). Any dul:concept that is linked to a ground entity in an *episode*, e.g. via the dul:classifies relation, is defined in a *narrative* that describes the *episode* (12.17). Specifications of this relation further constrain the type of the construed dul:concept (12.18) and how the dul:concept is related to the *narrative* that defines it (12.19). Another specification of this relation further constrains the type of the narrativised dul:task (12.20) and how the dul:concept is related to the *narrative* that defines it (12.21). The formalisation of other sub-relations, i.e. *narrativizesRole* and *narrativizesTask*, is done analogously (Porzel 2021). The manifestation of a *narrative* (*MNAR*) is, in our view, a dul:situation that *satisfies* (*SAT*) the *narratives* describing the episodic entities that are included in the dul:situation (12.22). More concretely, it is a dul:situation where a dul:agent executes the dul:task defined by the *episode* by following a dul:plan (which is another type of dul:description) involving dul:physicalObjects playing certain dul:roles and dul:regions setting specific dul:parameters for that execution. Hence, dul:situations in which *narratives* are manifested also satisfy the dul:plan that the dul:agent executes (12.23).

$$MNAR(x) \rightarrow \exists! y(SAT(x, y) \rightarrow Narrative(x)) \tag{12.22}$$

$$MNAR(x) \rightarrow \exists! y(SAT(x, y) \rightarrow Plan(x)) \tag{12.23}$$

A formalisation of dul:plans that describe dul:tasks evoked by *episodes* is part of the existing DUL ontology. What makes a narrative a special type of description is then, among other things, that it assumes a specific *perspective*. As this needs to be included in a corresponding model, we propose extending the work from psycholinguistics (Herrmann and Grabowski 1994) to the narrative domain as follows:

$$Perspective(x) \rightarrow Abstract(x) \tag{12.24}$$

$$HASO(x, y) \rightarrow Perspective(x) \wedge Origo(y) \tag{12.25}$$

$$Origo(x) \rightarrow Role(x) \tag{12.26}$$

$$Egocentric(x) \rightarrow Perspective(x) \wedge Origo(Speaker) \tag{12.27}$$

$$Allocentric(x) \rightarrow Perspective(x) \wedge Origo(\neg Speaker) \tag{12.28}$$

Lastly, we can connect *narratives* with their assumed *perspective* by postulation of the relation of having a point of view (HAS_{PoV}) that holds between *narratives* and *perspectives* (12.29):

$$\texttt{Narrative}(x) \rightarrow \forall y(HAS_{PoV}(y, x) \rightarrow \texttt{Perspective}(y)) \tag{12.29}$$

In the next section, we show how such a model can be employed to serve as an intermediary representation that facilitates turning a knowledge graph—that represents an event—into a linguistically realised narrative, which tells a story about this event.

12.4 From Knowledge Graphs to Linguistic Expressions

The method we showcase in this work features two main steps that follow one another:

- Firstly, a formal representation of a narrative as triples is constructed based on an existing knowledge graph of a given event (the *fabula*). For this, the events and their participants as defined in the *fabula* are linked to different narratives that describe these events and, consequently, can consider them from different perspectives. By filtering based on a given perspective and choosing which events and participants to include, the narrative itself is constructed. The result of this filtering then gives what in narratology is usually referred to as the *plot* as introduced above (Meghini et al. 2021).
- Secondly, the narrative—in the form of said triples—is converted into a semantic specification of the text to be produced, which is then converted into text using a natural language generation system based on KPML and systemic functional grammar, as indicated above (Bateman 1997).

In the following, we provide illustrative details of our approach for turning knowledge graphs into natural language text using examples from the domain of current world affairs. Figure 12.3 shows an overview of the steps necessary to go from the underlying knowledge graph to the generated natural language text,

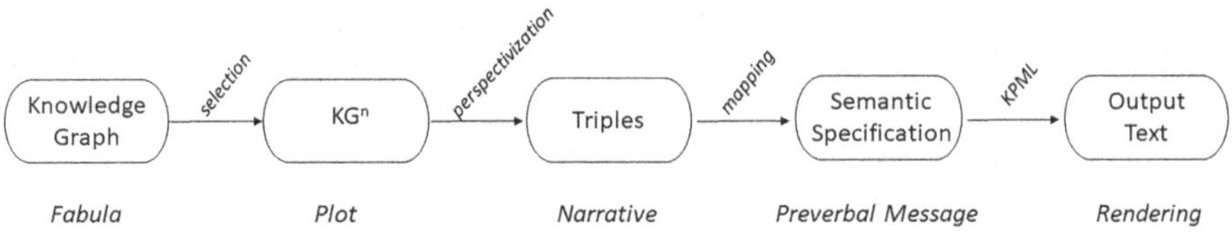

Fig. 12.3 The steps of converting a fragment of a Fabula into narrativised natural language text

which describes a chosen event from a specific narrative perspective. The following subsections set out the successive layers of this process.

12.4.1 Fabula

The underlying knowledge graph, which does not explicitly encode any kind of perspective on the given events, is what we consider as the *fabula*. In the future, we aim to use existing resources, such as EventKG (Gottschalk and Demidova 2019), as the basis for constructing formal narratives of events. However, since existing resources include large amounts of knowledge but also often miss important relations, this would still require a significant amount of manual work. Consequently, we present this case study on small-scale examples, for which we constructed the base knowledge graphs manually using the information available in EventKG.

12.4.2 Plot and Narrative

The selected knowledge graph is expanded into narratives that can assume different perspectives with respective EVENTS that are mapped to TASKS and PARTICIPANTS that are mapped to their respective ROLES. This content selection and filtering based on perspective produces a *narrative specification* represented as triples. Figure 12.5 gives an overview as to what sorts of entities and relationships can appear in narrative specification triples.

12.4.3 Semantic Specification

The narrative specification triples are mapped to semantic structures such as discourse relations between events (motivation, explanation, concession), action specifications or object descriptions. Assembling these structures then results in a semantic specification, which is expressed as a typed feature structure that represents its elements using concepts from an overarching linguistic ontology consisting of two modules: the Generalised Upper Model (GUM: Bateman 2022) and the Upper Interaction Ontology (Ross 2011, 120–126).

12.4.4 Tactical Generation

Once a semantic specification exists, it is transferred to the language generation software KPML to produce finely controlled natural language text in a selected language.[2]

12.4.5 Narrativisation of Knowledge Graphs

Figure 12.4 shows a very small set of events and the participants included in them. In prior work (Spillner et al. 2022), a large corpus of tweets concerning the ongoing war in Ukraine was collected. This data was analysed in order to reconstruct different narratives surrounding this conflict, which have been shared on social media. Since this existing dataset includes texts written by humans describing these events from many different perspectives, we used these events as a starting point that is formalised for this example. A listing of such knowledge graphs representing specific events is provided in Table 12.1; these are then construable into the narratives depicted in Fig. 12.4.

As proposed in the formal model adopted for representing narratives outlined above (Porzel and Cangalovic 2020), narratives define TASKS that are executed in the given EVENTS and the ROLES that the events' participants can take. Based on this formal theory of narratives, we consider several ways in which a neutral knowledge graph can be narrativised:

12.4.5.1 Event—Task

An event that exists in the KG can be construed in different tasks: for example, E1 in Fig. 12.4 represents the invasion of Ukraine by Russia, which, depending on the speaker's point of view, might be characterised, e.g. as an invasion, the launch of a special military operation or even as a liberation.

[2] The general KPML system and a collection of grammars are available at the URL purl.org/net/kpml; the grammars used for the experiments reported here are slightly extended versions of the Nigel grammar of English and a grammar for Spanish developed as part of a doctoral project by Juan Rafael Zamorano Mansilla funded by Bremen University, 2000–2003.

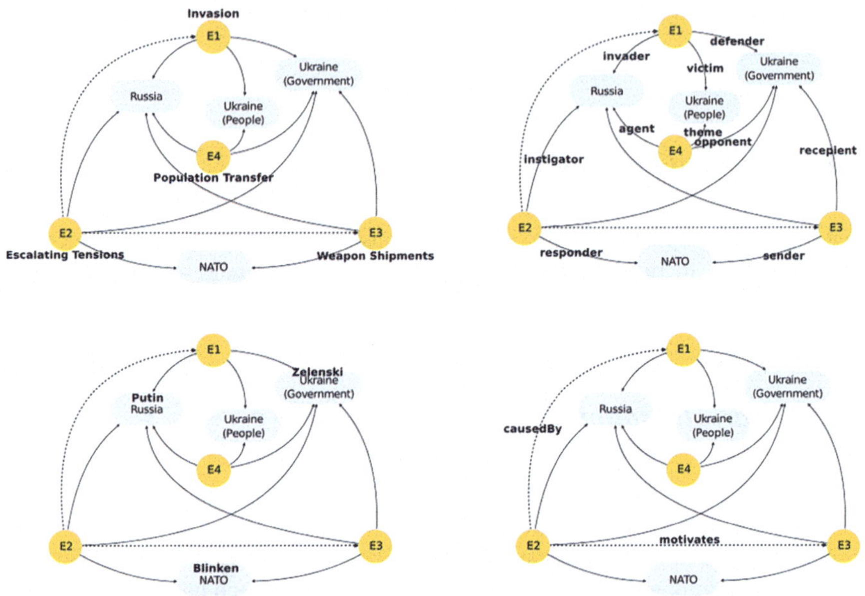

Fig. 12.4 Ways to narrativise a given event knowledge graph. Clockwise from top left: adding event type; adding roles to participants; adding relationships between events; selecting terminology/part-whole substitutions

Table 12.1 Event knowledge graph triples functioning as input specifications for the narrativisation process

Event Knowledge Graphs (Input)
(E1, hasParticipant, Ukraine(people))
(E1, hasParticipant, Ukraine(government))
(E1, hasParticipant, Russia)
(E2, hasParticipant, Ukraine(government))
(E2, hasParticipant, NATO)
(E2, hasParticipant, Russia)
(E3, hasParticipant, NATO)
(E3, hasParticipant, Ukraine(government))
(E3, hasParticipant, Russia)
(E4, hasParticipant, Ukraine(people))
(E4, hasParticipant, Ukraine(government))
(E4, hasParticipant, Russia)

12.4.5.2 Participant—Role

A narrative also defines at least one role in the task that is taken by the participants of the event. Thus, different narratives might include very different kinds of roles, e.g. an invader *vs.* a liberator, as well as having different participants take those roles, e.g. in E2, the escalation of the Russia-Ukraine crisis, the role of the agent causing the escalation might be taken by Russia, by NATO or by other participants, depending on the point of view being constructed by the narrative.

12.4.5.3 Terminology

Terms used to refer to events or their participants can impact the surrounding narratives in different ways.

Firstly, collections of knowledge graphs, such as provided by EventKG, which combine information from several distinct sources, provide different terms for the events themselves by combining events from different sources through the "sameAs" relationship. For example, depending on the source, the ongoing war in Ukraine is referred to as "Ukrainian War", "Russo-Ukrainian War" or "Russian military intervention in Ukraine", and the annexation of Crimea can be found under "Annexation of Crimea by Russian Federation", "Crimean Crisis 2014" or "armed political conflict surrounding Crimea". Additionally, alternative names are also provided, including "Secession and Incorporation of Crimea" as referring to the annexation of Crimea or "Russian Spring", which refers to the pro-Russian unrest in Ukraine in 2013.

Secondly, the events included in EventKG also have a "type"—thus, the Ukrainian war might be seen not only as a war but also as an armed conflict, a military conflict or an "intervention", while the aforementioned unrest in 2014 is defined, e.g. as "secession", "protest", "civil disorder" or "insurgency".

Thirdly, terminology can be used to characterise the participants in the event. This ranges from smaller distinctions such as the usage of a person's full name or title (e.g. "Putin" *vs.* "President Vladimir Putin" *vs.* "dictator Putin") to terms that also cast participants in very specific roles regarding the task defined in the narrative: an example of this in the Twitter conversation surrounding the war in Ukraine is the casting of its invasion as a liberation—however, not only is the invasion itself characterised as aiming to liberate the country in general, but the oppressor it is to be liberated from is "the Nazis".

This term can be found in different contexts. In tweets talking about "liberation from the Nazis", the latter usually refers more generally to the Ukrainian government, which thus takes the role of oppressor in the liberation task. However, when this narrative is challenged, the term is defended by refocusing it on other groups, usually the Azov regiment of the Ukrainian Army (Spillner et al. 2022).

12.4.5.4 Relationships Between Events

In EventKG, the events themselves are connected only through sub-event relationships, but not in terms of their causality or other relations. The combination of several events in textual expressions can add additional connections that are by and large contingent on the narrative perspective: an event might be caused by a different one, be motivated by another, or happen in spite of another event that opposes it.

For example, depending on the perspective of the speaker, the invasion of Ukraine by Russia (E1) might be caused by continuously escalating tensions (E2) to which Russia is reacting, or it might be started by Russia with the goal of further escalating the existing crisis.

12.4.5.5 Perspective Filtering

Consequently, through the steps described above, a perspective has been chosen that primarily involves the selection of a frame-giving conceptualisation of the event. In the terminology of SOMA, this corresponds to selecting a TASK that *classifies* an EVENT. Analogous to frame entities found in FrameNet (Baker et al. 1998), each specific frame contains a frame-specific configuration of *roles*, e.g. invading features invaders and invaded entities whereas liberating involves the liberated and the liberators. Correspondingly, one and the same entity can be endowed with a positive valence, e.g. the liberator, or a negative one, e.g. the invader, albeit both are instantiated by the same entity in the knowledge graph. A set of sample configurations was given in Fig. 12.4.

12.4.6 From Narrative Specification to Texts

As previously mentioned, for our purposes here, a narrative specification is a collection of triples asserting that some events happened and had various entities as participants. The events may have been connected to each other by relations such as:

- opposition—one event should have prevented another but did not
- motivation—one event happened because its agentive participants desired some other event to happen
- explanation—one event happened because another one happened

A summary of the available relationships and entity types is given in Fig. 12.5.

As described above, for the final part of the language generation process, i.e. realisation of surface forms, we use KPML (Bateman 1997). KPML does not accept triples as inputs but rather semantic specifications (semspecs), which are represented as typed feature structures with types drawn from the defined linguistic ontology. A very simple example of such a semantic specification for the nominal phrase "the crisis" is:

```
(OBJ_0b4ax4 / |Object| :LEX CRISIS :determiner the)
```

A semspec is headed by an identifier for the semantic entity for which a natural language expression is to be generated, followed by its linguistic ontological characterisation in terms of a semantic type defined in the Generalised Upper Model or the Upper Interaction Ontology, a set of participants or circumstantial information

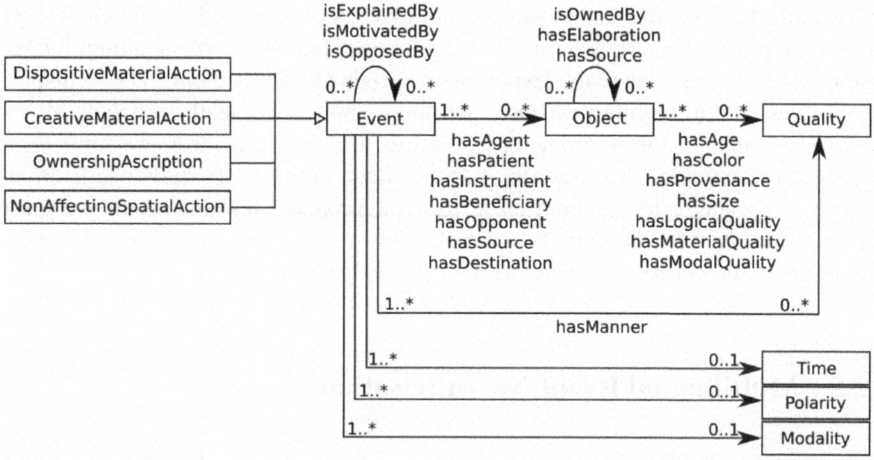

Fig. 12.5 The ontological schema of a narrative specification. The main entities are events, which can be of different types and which have objects as participants. Events and objects can have qualities

further characterising the semantics of the entity being realised and various further lexico-semantic specifications such as a lexical item to use in the realisation, the form of determiners to be used for referring to objects, flags concerning textual identifiability and many more. Semspecs typically include further semspecs recursively or refer to them via identifiers. In particular, a semspec for an event or object will typically include semspecs for entities playing some role for that event or object. Lexical information may be given directly, as shown in our example, but is more often associated with concepts so that explicit specifications in semspecs are no longer necessary. We maintain the explicit specifications for the purposes of the current discussion, however, so that the semspecs are locally complete.

KPML is also particularly oriented to multilingual natural language generation and description in that all levels of linguistic specification, from lexical entries to semantics, may be conditionalised according to the language varieties for which they hold; this supports strong notions of resource sharing across languages (Bateman 2000), which we may draw on more systematically in the future. In the examples discussed below, two largely distinct grammars are used with respect to the shared linguistic ontology.

The conversion from a set of triples—the narrative specification—to a semspec then makes use of the compositionality of KPML semspecs. That is, to generate a semantics specification for a relationship between events, it is enough to generate a semspec for each event participating in the relation and then generate a semspec for the relation according to an appropriate template, which is then filled in with the semspecs for the participating events. Likewise, to generate a semspec for an event is to fill in an appropriate template with the semspecs produced for the participating objects and so on.

In Table 12.2, we list some example narrative specs produced for the knowledge graphs augmented as shown in Fig. 12.4 above and the corresponding KPML outputs, i.e. natural language expressions, produced from them. The semspecs themselves tend to be more verbose than those shown here as they include all the necessary semantic information for fine-grained narrative control; we omit these details here, therefore, for reasons of space. Each entity in the full specifications also has associated a lexical item via a "hasLex" property: however, again for space reasons, we do not list these in the examples shown as for these examples, the entity names are sufficient for the reader to infer the lexical item.

12.5 Multilingual Event Narrativisation

The semantic specifications produced allow us to "translate" the language output into specific languages, such as Spanish, on the surface level. At the end of the pipeline, lexicographic and morpho-syntactic information in the form of a lexicon has to be specified, which is then converted into natural language with the help of a grammar specified for the language at hand and the pre-existing concepts from the Generalised Upper Model. Although the semantic specifications produced tend to be relatively language independent due to their focus on the generic linguistic ontology employed, there can always be language-specific differences in how events and objects are to be construed that require adaptation. This is a very broad issue that has received considerable attention in both multilingual natural language generation and machine translation: in general, it is not possible to locate any level of description that can be guaranteed to be invariant across languages (Bateman et al. 1991; Delin et al. 1993; Vander Linden and Scott 1995).

Due to language-specific differences, therefore, it is sometimes necessary to adapt the semantic specifications we produce. One example is the sentence "Russia is abducting people from Ukraine", which translates into Spanish as "Rusia está secuestrando a personas en Ucrania". The semantic difference lies in the reference to "Ukraine". In the English example, the country is displayed as a location with the inherent notion of a *source* indicated by the preposition "from", whereas in the Spanish sentence, it is expressed as a location without any direction. This is due to Spanish being a *verb-framed* language, where the component of the *path* is already encoded semantically into the verb, in contrast to *satellite* languages, such as English, where the information regarding the path is expressed grammatically via a corresponding circumstantial prepositional phrase of direction (Talmy 2000). Such broad-scale differences are themselves often systematic, however, and so can be approached in a principled manner via suitable conditionalisations of narrativisation templates prior to reaching the final tactical generation phase.

Table 12.2 Narrativised events and their English and Spanish surface realisations

Narrative specification	English output	Spanish output
('construedAs', 'invade', 'dispmatact') ('hasAgent', 'invade', 'russia') ('hasPatient', 'invade', 'ukraine') ('hasTime', 'invade', 'past')	Russia invaded Ukraine.	Rusia invadió Ucraina.
('construedAs', 'liberate', 'dispmatact') ('hasAgent', 'liberate', 'russia') ('hasPatient', 'liberate', 'ukraine') ('hasOpponent', 'liberate', 'nazis') ('hasTime', 'liberate', 'present–continuous') ('hasDeterminer', 'nazis', 'the')	Russia is liberating Ukraine in spite of the Nazis.	Rusia está liberando Ucraina a pesar de los nazis.
('construedAs', 'liberate', 'dispmatact') ('hasAgent', 'liberate', 'russia') ('hasPatient', 'liberate', 'ukraine') ('hasSource', 'liberate', 'nazis') ('hasTime', 'liberate', 'present–continuous') ('hasDeterminer', 'nazis', 'the')	Russia is liberating Ukraine from the Nazis.	Rusia está liberando Ucraina de los nazis.
('construedAs', 'launch', 'dispmatact') ('hasAgent', 'launch', 'russia') ('hasPatient', 'launch', 'operation') ('hasDestination', 'launch', 'ukraine') ('hasSize','operation','special') ('hasMatQuality','operation','military') ('hasTime', 'launch', 'past') ('hasDeterminer','operation','a')	Russia launched a special military operation to Ukraine.	Rusia lanzó una operación militar contra Ucraina.
('construedAs', 'escalate', 'dispmatact') ('hasAgent', 'escalate', 'russia') ('hasPatient', 'escalate', 'crisis') ('hasQuality', 'crisis', 'ongoing') ('hasTime', 'escalate', 'present') ('hasModality', 'escalate', 'can') ('isMotivatedBy', 'invade', 'escalate')	Russia invaded Ukraine, so that Russia can escalate the ongoing crisis.	Rusia invadió Ucraina, para que Rusia pueda agravar la crisis actual.
('construedAs', 'escalate', 'dispmatact') ('hasAgent', 'escalate', 'nato') ('hasPatient', 'escalate', 'crisis') ('hasQuality', 'crisis', 'ongoing') ('hasTime', 'escalate', 'past') ('isExplainedBy', 'liberate', 'escalate')	Russia liberated Ukraine, because NATO escalated the ongoing crisis.	Rusia liberó Ucraina porque la OTAN agravó la crisis actual.

(continued)

Table 12.2 (continued)

Narrative specification	English output	Spanish output
('construedAs', 'escalate', 'dispmatact') ('hasAgent', 'escalate', 'nato') ('hasPatient', 'escalate', 'crisis') ('hasQuality', 'crisis', 'ongoing') ('hasTime', 'escalate', 'past') ('isExplainedBy', 'liberate', 'escalate')	Russia liberated Ukraine, because NATO escalated the ongoing crisis.	Rusia liberó Ucraina porque la OTAN agravó la crisis actual.
('construedAs', 'leave', 'nonaffspat') ('hasAgent', 'leave', 'people') ('hasSource', 'leave', 'ukraine') ('hasDestination', 'leave', 'russia') ('hasTime', 'leave', 'present-continuous') ('hasDeterminer', 'people', 'the')	The people are leaving from Ukraine to Russia.	La gente se va de Ucrania a Rusia.
('construedAs', 'abduct', 'dispmatact') ('hasAgent', 'abduct', 'russia') ('hasSource', 'abduct', 'ukraine') ('hasPatient', 'abduct', 'people') ('hasTime', 'abduct', 'present-continuous')	Russia is abducting people from Ukraine.	Rusia está secuestrando a personas en Ucrania.

12.6 Conclusion and Future Work

In this work, we have introduced a pipeline for turning knowledge graphs into natural language texts in different languages, i.e. English and Spanish for the moment. These event descriptions assume a specific perspective on the propositional content given by the respective knowledge graphs. For this, a formal model of narratives provides the ontological foundation for representing the intermediate perspective-specific representation. It should also be noted that not only do natural language expressions contain biases and value judgements, but ontological models and knowledge graphs can feature biases imported by their designers as well (Keet 2021). In this approach, however, the perspective is explicitly modelled as an integral part of the narratives at hand. This feature enables us to produce narrative-specific descriptions of a given knowledge graph that, as opposed to, e.g. tweets in the wild, makes the underlying perspective openly available for inspection via the narrative specification.

Our approach, however, still needs substantial work to make it scalable and generally applicable to any event representation, i.e. knowledge graph, available in some collection and to any language. For this, ways have to be found for:

- dealing with missing information in knowledge graphs, e.g. in EventKG, the person Louis XVI is not given as a participant of his own beheading.

- automatic creation of hermeneutic filters for mapping the events contained in a knowledge graph to possible interpretations thereof. One approach for this could employ generic ontology design patterns (Krieg-Brückner et al. 2020).
- adding register, i.e. variation according to use, to the natural language generation output (Bateman and Paris 1991). Lexical choices can also express valence and ideological positioning about some entity, e.g. the difference between referring to an atom bomb as a tactical and a nuclear weapon carries some opinion with it (Hovy 1990).
- adding more languages to test further the degree of language independence of the semantic specifications and narrative model.

To evaluate the output of our system, a type of *Bleu Score* evaluation could be undertaken straightforwardly (Papineni et al. 2002). However, since it is not yet the intent of this work to create textual narratives for human readers but rather to explore the kind of variation that can be productively controlled in our narrativisation process, an evaluation of their structural similarity to real conflictual narratives (Spillner et al. 2022) would arguably be more appropriate at this stage. For this, some annotation-based metrics, e.g. Cohen's Kappa (Cohen 1960), could be employed to compare semantic annotations of real and generated narratives.

Our hope is that this work will constitute the beginning of a research effort that complements current efforts that focus on going from natural language expressions to formal specifications—either narratives or knowledge graphs—by looking at the reverse direction. Ultimately, the long-term research effort behind this undertaking concerns an improvement of our understanding of narrative mechanics, i.e. how descriptions of events are constructed, manipulated and finally expressed as natural language utterances with a fully explicit record of those utterances' perspectivisation of the events addressed.

Acknowledgments The research reported in this chapter has been supported by the FET-Open Project 951846 MUHAI—"Meaning and Understanding for Human-centric AI"—funded by the EU Program Horizon 2020 (http://www.muhai.org/) as well as the German Research Foundation DFG, as part of Collaborative Research Centre (Sonderforschungsbereich) 1320 EASE—"Everyday Activity Science and Engineering"—University of Bremen (http://www.ease-crc.org/). The research was conducted in sub-project *P01 Embodied semantics for the language of action and change: Combining analysis, reasoning and simulation.*

References

Baker CF, Fillmore CJ, Lowe JB (1998) The Berkeley FrameNet Project. In: Proceedings of the 36th annual meeting of the association for computational linguistics and 17th international conference on computational linguistics, vol 1. Association for Computational Linguistics, USA, ACL '98/COLING '98, pp 86–90. https://doi.org/10.3115/980845.980860, https://doi.org/10.3115/980845.980860

Bateman JA (1997) Enabling technology for multilingual natural language generation: the KPML development environment. J Nat Lang Eng 3:15–55

Bateman JA (2000) Multilinguality and multifunctionality in linguistic description - and some possible applications. Sprachtypologie und Universalienforschung (STUF) 53(2):131–154

Bateman JA (2022) GUM: the generalized upper model. Appl Ontol 17(1):107–141. https://doi.org/10.3233/AO-210258

Bateman JA, Paris CL (1991) Constraining the development of lexicogrammatical resources during text generation: Towards a computational instantiation of register theory. In: Ventola E (ed) Recent systemic and other views on language, Mouton, Amsterdam, pp 81–106

Bateman JA, Matthiessen CMIM, Nanri K, Zeng L (1991) The re-use of linguistic resources across languages in multilingual generation components. In: Proceedings of the 1991 international joint conference on artificial intelligence, Sydney, vol 2, pp 966–971. Australia, Morgan Kaufmann Publishers, San Francisco

Beßler D, Porzel R, Pomarlan M, Vyas A, Höffner S, Beetz M, Malaka R, Bateman J (2020) Foundations of the Socio-physical Model of Activities (SOMA) for autonomous robotic agents. CoRR. https://arxiv.org/abs/2011.11972

Beßler D, Porzel R, Pomarlan M, Beetz M, Malaka R, Bateman J (2020) A formal model of affordances for flexible robotic task execution. In: European Conference on Artificial Intelligence (ECAI)

Blin I (2022) Building narrative structures from knowledge graphs. In: European semantic web conference. Springer, Berlin, pp 234–251

Cohen J (1960) A coefficient of agreement for nominal scales. Educ Psychol Meas 20(1):37

Delin JL, Scott D, Hartley T (1993) Knowledge, intention, rhetoric: levels of variation in multilingual instructions. In: Rambow O (ed) Intentionality and structure in discourse relations. Association for Computational Linguistics, pp 7–10 (Proceedings of a Workshop sponsored by the Special Interest Group on Generation, 21 June, 1993, Columbus, OH)

Eger M, Barot C, Young RM (2015) Impulse: a formal characterization of story. In: Finlayson MA, Miller B, Lieto A, Ronfard R (eds) Proceedings of 6th workshop on computational models of narrative (CMN 2015), pp 45–53

Elson DK (2004) Categorization of narrative semantics for use in generative multidocument summarization. In: Belz A, Evans R, Piwek P (eds) Natural language generation: third international conference (INLG 2004), no. 3123. Lecture notes in artificial intelligence. Springer, Berlin, pp 192–197

Evans R, Hernández-Orallo J, Welbl J, Kohli P, Sergot M (2021) Making sense of sensory input. Artif Intell 293:103438

Gangemi A, Mika P (2003) Understanding the semantic web through descriptions and situations. In: On the move to meaningful internet systems 2003: CoopIS, DOA, and ODBASE - OTM confederated international conferences, CoopIS, DOA, and ODBASE 2003, Catania, Sicily, November 3–7, 2003, pp 689–706

Gottschalk S, Demidova E (2019) EventKG–the hub of event knowledge on the web–and biographical timeline generation. Semantic Web 10(6):1039–1070

Guarino N, Welty C (2004) An overview of OntoClean. In: Staab S, Studer R (eds) Handbook on ontologies. Springer, Heidelberg, pp 151–171

Herrmann T, Grabowski J (1994) Sprechen: Psychologie der Sprachproduktion. Spektrum, Akad. Verl., Heidelberg [u.a.]. https://madoc.bib.uni-mannheim.de/17911/

Hovy EH (1990) Pragmatics and natural language generation. Artif Intell 43(2):153–197

Jasinskaja K, Roßdeutscher A (2012) Through narrative planning towards the preverbal message. In: Benz A, Kühnlein P, Stede M (eds) Constraints in discourse 3: representing and inferring discourse structure. John Benjamins, Amsterdam, pp 45–76

Keet CM (2021) An exploration into cognitive bias in ontologies. In: Sanfilippo EM, Kutz O, Troquard N, Hahmann T, Masolo C, Hoehndorf R, Vita R, Hedblom MM, Righetti G, Sormaz D, Terkaj W, Sales TP, de Cesare S, Gailly F, Guizzardi G, Lycett M, Partridge C, Pastor O, Beßler D, Borgo S, Diab M, Gangemi A, Alarcos AO, Pomarlan M, Porzel R, Jansen L, Brochhausen M, Porello D, Garbacz P, Seppälä S, Grüninger M, Vizedom A, Dooley DM, Warren R, Küçük-McGinty H, Lange M, Algergawy A, Karam N, Klan F, Michel F, Rosati I (eds) Proceedings of the joint ontology workshops 2021 episode VII: The Bolzano summer of knowledge co-located with the 12th international conference on formal ontology in information systems (FOIS 2021), and the 12th international conference on biomedical ontologies (ICBO 2021), Bolzano, September 11–18, 2021, CEUR-WS.org, CEUR Workshop Proceedings, vol 2969. http://ceur-ws.org/Vol-2969/paper38-CAOS.pdf

Krieg-Brückner B, Mossakowski T, Codescu M (2020) Generic ontology design patterns: roles and change over time. https://doi.org/10.48550/ARXIV.2011.09353, https://arxiv.org/abs/2011.09353

Kroll H, Nagel D, Balke WT (2020) Modeling narrative structures in logical overlays on top of knowledge repositories. In: Dobbie G, Frank U, Kappel G, Liddle SW, Mayr HC (eds) Concpetual modeling. Springer, Heidelberg, pp 19–33

Kurisinkel LJ, Chen NF (2021) Graph to coherent text: passage generation from knowledge graphs by exploiting edge representations in sentential contexts. In: Proceedings of the AAAI workshops on commonsense knowledge graphs (CSKGs), Virtual

Lakoff G, Narayanan S (2010) Towards a computational model of narrative. In: Proceedings of the AAAI fall symposium on computational models of narrative. American Association for Artificial Intelligence, Technical Report FS-10-04

Lehnert WG (1982) Plot units: a narrative summarization strategy. In: Lehnert WG, Ringle MH (eds) Strategies for natural language processing. Lawrence Erlbaum Associates, Hillsdale, pp 375–414

Mani I (2012) Computational modeling of narrative. Synth Lect Hum Lang Technol 5(3):1–142

Marcheggiani D, Perez-Beltrachini L (2018) Deep graph convolutional encoders for structured data to text generation. In: Proceedings of the 11th international conference on natural language generation. Association for Computational Linguistics, Tilburg University, pp 1–9. https://doi.org/10.18653/v1/W18-6501, https://aclanthology.org/W18-6501

Masolo C, Borgo S, Gangemi A, Guarino N, Oltramari A (2003) WonderWeb deliverable D18 ontology library (final). Technical report, IST Project 2001-33052 WonderWeb: Ontology Infrastructure for the Semantic Web

Meehan JR (1980) The Metanovel: writing stories by computer. Garland

Meghini C, Bartalesi V, Metilli D (2021) Representing narratives in digital libraries: the narrative ontology. Semant Web J 12(11):1–24

Papineni K, Roukos S, Ward T, Zhu WJ (2002) Bleu: a method for automatic evaluation of machine translation. In: Proceedings of the 40th annual meeting on association for computational linguistics. Association for Computational Linguistics, Stroudsburg, pp 311–318

Peinado F, Pablo Gervás BDA (2004) A description logic ontology for fairy tale generation. In: Proceedings of the workshop on language resources for linguistic creativity, LREC, pp 56–51

Porzel R (2021) On formalizing narratives. In: Proceedings of the joint ontology workshops 2021 episode VII: The Bolzano summer of knowledge co-located with the 12th international conference on formal ontology in information systems (FOIS 2021), and the 12th international conference on biomedical ontologies (ICBO 2021), CEUR-WS.org, CEUR Workshop proceedings, vol 2969. https://ceur-ws.org/Vol-2969/paper31-CAOS.pdf

Porzel R, Cangalovic VS (2020) What say you: an ontological representation of imperative meaning for human-robot interaction. In: Proceedings of the JOWO – ontology workshops, Bolzano, http://ceur-ws.org/Vol-2708/robontics4.pdf

Reiter E (1994) Has a consensus NL generation architecture appeared, and is it psychologically plausible? In: McDonald D, Meteer M (eds) Proceedings of the 7th international workshop on natural language generation (INLGW '94), Kennebunkport, pp 163–170

Reiter E, Dale R (2000) Building natural language generation systems. Natural language processing. Cambridge University Press, Cambridge, https://doi.org/10.2277/052102451X

Riedl MO, Young RM (2010) Narrative planning: balancing plot and character. J Artif Intell Res 39:217–268

Ross RJ (2011) Situated dialogue systems: agency & spatial meaning in task-oriented dialogue. CreateSpace Independent Publishing Platform, Scotts Valley

Spillner L, Santagiustina C, Mildner T, Porzel R (2022) Towards conflictual narrative mechanics. In: Proceedings of the IJCAI/ECAI workshop on semantic techniques for narrative-based understanding

Swartjes I, Theune M (2006) A fabula model for emergent narrative. In: Göbel S, Malkewitz R, Iurgel I (eds) Technologies for interactive digital storytelling and entertainment, no. 4326. LNCS, Springer, Berlin, pp 49–60

Szilas N (2007) A computational model of an intelligent narrator for interactive narratives. Appl Artif Intell 21(8):753–801

Talmy L (2000) Toward a cognitive semantics, volume 1: concept structuring systems. The MIT Press, Cambridge, https://doi.org/10.7551/mitpress/6847.001.0001, https://doi.org/10.7551/mitpress/6847.001.0001

Vander Linden K, Scott D (1995) Raising the interlingual ceiling with multilingual text generation. In: Proceedings of the IJCAI workshop in multilingual text generation (International joint conference on artificial intelligence) 1995, Montréal, pp 95–101

Welty C, Andersen W (2005) Towards OntoClean 2.0: a framework for rigidity. Appl Ontol 1(1):107–116

Winer D (2014) Review of ontology based storytelling devices. In: Dershowitz N, Nissan E (eds) Language, culture, computation: computing of the humanities, law, and narratives, no. 8002, LNCS. Springer, Berlin, pp 394–405

Zarri GP (2009) Representation and management of narrative information - theoretical principles and implementation. Advanced information and knowledge processing. Springer, Berlin. https://doi.org/10.1007/978-1-84800-078-0

Chapter 13
Outlook

Jane Winters

A number of the chapters in this volume consider the temporality of events and how this affects the ways in which they are encountered, analysed and discussed online. The CLEOPATRA project has its own very specific temporality. It was funded for 4.5 years from January 2019 to June 2023, although it was many months in the planning and evaluation beforehand and now has an extended afterlife (as evidenced by this book). Members of the project team joined and left at different times, with some being part of it from the start and others joining for only a few months. That is the normal trajectory of a research project, but the particular timing of CLEOPATRA has shaped the research that was undertaken in unprecedented ways. First, of course, the project took place against the background of a global pandemic. Just over a year after the project started, in March 2020, pandemic lockdowns came into force around the world. Like so many others, the project team had to adapt to new ways of working. Tools and platforms for online collaboration were integrated into workflows, different kinds of seminars, conferences and public engagement events were organised, and data sharing and publication began to be reimagined. 'Pandemic projects' like CLEOPATRA have been forced to pioneer new modes of research and practice, and some of this innovation will persist and become embedded in large-scale, interdisciplinary, international projects in the future.

The research presented in this volume relies substantially on the availability of open data, for example Wikidata, Wikipedia, DBpedia and YAGO. It also, like a great deal of research in the computer sciences, social sciences and humanities, made use of data accessed via the open Twitter API. This allowed the investigation of multilingual discourse concerned with online hate speech, climate change, the

J. Winters (✉)
School of Advanced Study, University of London, London, UK
e-mail: jane.winters@sas.ac.uk

I. Marenzi et al. (eds.), *Event Analytics across Languages and Communities*,
https://doi.org/10.1007/978-3-031-64451-1_13

COVID-19 pandemic and the impact of technology. In October 2022, Elon Musk concluded his acquisition of Twitter (subsequently rebranded as X), and free access to its API was closed in February of the following year. Data collection for the project had long since been completed, but that work of data collection and analysis could not now be carried out. The closure of the free API not only restricts access to data, it also immediately makes obsolete any tools and services that were built on top of the API. The relationship between academic researchers and commercial data owners is one that needs to be more critically evaluated in years to come. If not, there will be significant—and damaging—results for industry, economy and society. Commitments to Findable, Accessible, Interoperable and Reusable (FAIR) data are at the heart of research, but this cannot be secured without safeguards, guarantees and new approaches to collaboration between academia and industry.

The third major change that has occurred since the start of the CLEOPATRA project, as noted in the introduction to this volume, is the rapid advance in generative Artificial Intelligence (AI) and the growing pervasiveness of Large Language Models (LLMs). General public awareness of these technologies has also taken a significant leap forward, with media coverage of the hallucinations of ChatGPT, the roll out of AI-integrated search by Bing and Google, and perhaps most pertinently, the existence of simple and intuitive user interfaces. In addition, while there is likely to be some artificially generated material in the data collected by the project team—it is impossible to rule out the inclusion of bots in a Twitter dataset, for example—the resources published by CLEOPATRA primarily represent human responses to events. The parameters have already changed, as generative AI is incorporated into a range of online media. In one sense, then, this book has rather unexpectedly become a snapshot of the state of the art in a pre-LLM era. The project has become interesting not just for its direct findings, but for its function as a key benchmark in multilingual research.

The work presented here has not, however, been superseded by generative AI and LLMs; rather, it is complementary and additive. First, the methods developed by the CLEOPATRA team embody critical approaches to data analysis that will be ever-more pressing as it becomes difficult, if not impossible, to distinguish between the human- and the computer-generated. The impact of LLMs in many sectors has led to a new emphasis on explainable AI, building on longstanding criticism of the 'black-boxing' of so many digital technologies. The meticulous documentation and transparency of data and method that characterises the research published in this book models an alternative way of proceeding, one that is open, explicable and trustworthy.

Questions around language equity and data access will persist. There is no evidence to suggest that the low-resourced languages which provided such an important focus for the project will remain anything other than low-resourced in the near to medium future. The skills to work with multilingual data, to compare the results of different tools and approaches in varying linguistic contexts, and to build tools and datasets that reflect the needs of minoritised groups worldwide will continue to be in demand. Finally, a number of the chapters in this volume have foregrounded the importance of narrative and storytelling for effective and meaningful engagement with data about events, with the role of the human and computational made explicit. There is one outlook in which this collaboration between human and machine becomes a process of genuine co-creation, with the unique value of both identified and acknowledged.

The manufacturer's authorised representative in the EU is Springer Nature Customer Service Centre GmbH, Europaplatz 3, 69115 Heidelberg, Germany. If you have any concerns regarding our products, please contact ProductSafety@springernature.com

Printed and bound by CPI Group (UK) Ltd, Croydon, CR0 4YY
30/09/2026
02225567-0012